Composite Materials
Fabrication Handbook #2

John Wanberg

Published by:
Wolfgang Publications Inc.
Stillwater, MN 55082
www.wolfpub.com

Legals

First published in 2010 by Wolfgang Publications Inc.,
PO Box 223, Stillwater MN 55082

ISBN number: 978-1-941064-64-1

Composite Materials Fabrication Handbook #2

Acknowledgements

My sincere, heartfelt appreciation goes out to several people who have been instrumental in making this second book possible. They include the following:

Charlotte, Eden, and Miles, who take my eccentric passion for composites in stride and offer so much wonderful support through my creative endeavors…

Metropolitan State College of Denver for providing access to lab space and personnel throughout these various professional development and research efforts…

The students of Metro State's Industrial Design department who give me the opportunity to teach composites, while constantly presenting interesting molding and fabrication quandaries that help keep my skills tuned and growing…

Plasticare, Inc. of Denver, Colorado (www.plasticareinc.com) for continued product and technical assistance…

Timothy Remus, for giving me the opportunity to continue sharing this exciting topic with other fabricators, do-it-yourselfers, and the innumerable creative souls out there.

Introduction

Composite Materials: Fabrication Handbook -2 expands upon its predecessor, *Composite Materials: Fabrication Handbook -1*. It explains the world of composites fabrication more in-depth by demonstrating several advanced, low-production methods of forming composite materials. These advanced techniques can help enable composites fabricators in creating better molding systems, improving a composite's surface quality, and further optimizing their composite creations by minimizing their weight, and maximizing their inherent stiffness and strength. The main focus of this book is to demonstrate methods for *forming* composites rather than to illuminate the engineering principles behind *designing* structural components. If readers are interested in the latter (which is a very lengthy topic, at best), engineering texts are available from several sources for that very purpose.

In the spirit of book one, many of the moldmaking and composite optimization methods shown in this book are well within the reach of an average builder, requiring only a modest investment in some specialized tools and materials to yield high-quality, professional results. Although researchers, engineers, and designers in the aerospace, auto racing, and sporting goods industries routinely expend significant effort to trim even a few ounces of weight from a composite, an advanced degree in engineering or manufacturing is not required to use many of their optimization methods. Several techniques are illuminated in this book, and include the following:

- "Composite tooling" fabrication methods (material originally left out of *Handbook-1*)
- Adding core materials for stiffer and stronger composite "sandwich" structures
- Compressing a laminate to improve its surface quality and fiber consolidation
- Using vacuum bagging to optimize a wet layup
- Laminating and processing "pre-preg"
- Employing expandable inserts or inflatable bladders in a mold system
- Performing vacuum-assisted resin transfer molding techniques to wet out a pre-form

Whatever your composite project may be, *Composite Materials: Fabrication Handbook - 2* is sure to shed some light on several helpful fabrication methods. With a little practice in these techniques, fabricators will be able to bring their composites ideas to life with much improved strength, weight-savings, and quality.

Chapter One

Designing Moldable Composite Parts

The Basic Principles for Designing Quality Molds and Components

CHAPTER INTRODUCTION

Molds are extremely important in the creation of composite parts. To fabricate them effectively, however, several things affecting the quality and usability of the mold must be considered. The various steps involved in designing molds for composite parts are covered briefly in this chapter.

DESIGNING MOLDABLE PARTS

One of the greatest benefits of building a mold for a composite project is the ability to make additional copies of your project for use down the road. This is especially helpful if you experience one of the following: your original part becomes damaged and needs replacement, you need several copies for testing purposes, you get the entrepre-

Good parts require good molds, like this body panel mold for a fiberglass concept car that has Class "A" surfaces.

neurial itch to make multiple parts to sell to others, or a myriad of other reasons. A mold is especially helpful in producing composites that require a smooth surface (or "aesthetic surface") with minimal finishing work—like when molding visible carbon fiber outer surfaces or when using a gel coat. Regardless of your particular needs, the cost and effort required to build a mold will quickly pay for itself if you ever need to make more than one copy of a composite part.

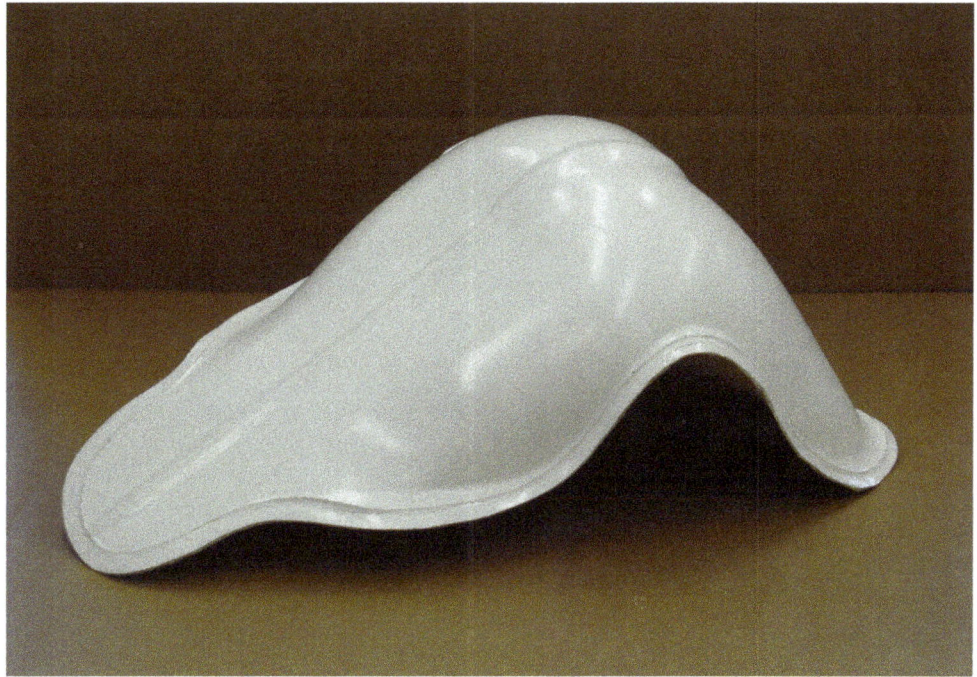

Example of a "male" mold.

GENERAL TERMINOLOGY

Before delving too far into the mold-making process, let's discuss some of the terminology common to the art. Laminates are always smoother where they come in direct contact with the mold surface. Therefore, a mold that is used to form a part with a smooth inner (or concave) surface is referred to as a "male" mold. Conversely, a mold that is used to form a smooth outer (or convex) surface on a part is called a "female" mold. To produce smooth surfaces on both sides of the laminate, certain mold systems, called "matched molds", are used. Therefore, fabricators choose a mold's geometry based on which side(s) they want to form the smooth or aesthetic surfaces.

Various types of hollow or otherwise complex shapes can be molded using a "mold core" or "insert"—or an internal molding component (somewhat similar to a

Example of a "female" mold.

Matched Mold System

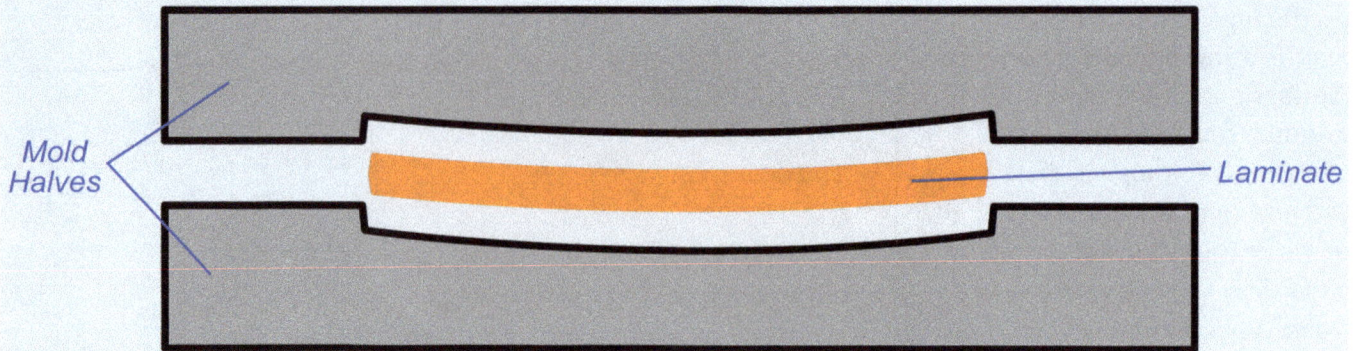

An example of matched molds.

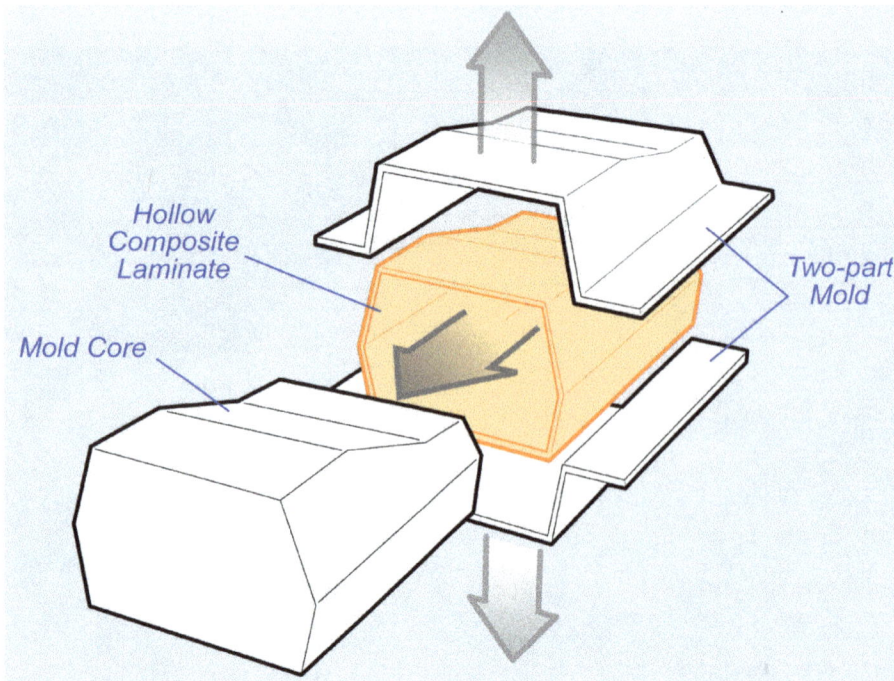

An example of a "mold core" or "insert".

mandrel) employed together with a female or male mold system. Some of these mold cores are considered "captive molds" in that they remain in the final molded composite—as with the hollow, thin-walled aluminum tanks around which filament-wound composites are formed to produce a pressure vessel. Some other types of mold cores can be released later, or destroyed or dissolved out of the part—as with a core made of plaster or styrene foam. Still other types of hollow shapes can be formed using expandable rubber inserts or inflatable bladders (as will be demonstrated later in this book) that can be removed after the composite cures. Lastly, when creating certain types of molds (such as composite tooling shown in chapter 2), the original part, prototype, or component around which the mold is created is usually referred to as the "pattern", "plug", "buck", or "master".

PART MOLDABILITY

In order to make an effective mold, it is important to understand what makes a part "moldable". An optimally moldable part is one that can be easily

8

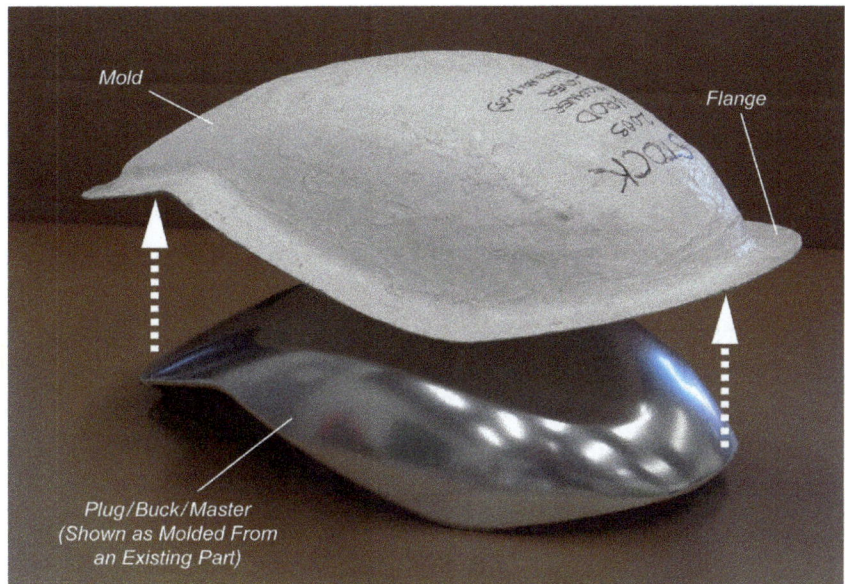

Relationship between an existing part and a mold produced from it.

removed from a mold while maintaining its intended shape without damaging it or the mold. However, not every part can be easily molded in composites, so a practical analysis of a part's shape, complexity, and surface texture is important to help determine what molding methods to use.

The first thing about a part's shape that will determine its moldability is its "draft"—or the angle of the part's sides in relation to the direction that it will be removed from the mold. A simple illustration of draft is found in the typical plastic ice cube tray in your freezer. Without angled sides, the brittle ice cubes formed in the tray would be difficult or impossible to remove from the tray—leaving us sipping warm drinks. One quick way to assess the draft on a part is to first look at it from a distance (preferably from several feet away) in the direction that it will be removed from a mold. If all the mold-contacting sides of the part are visible at the same time, then it has proper draft. As a general guideline, deep or narrow composite parts should be formed with at least 5 to 15 degrees of draft, while shallow or wide parts may release well with only 2 to 5 degrees of draft designed into them. A protractor or square can be used to assess the amount of draft on a mold. When a composite part must have a shape or

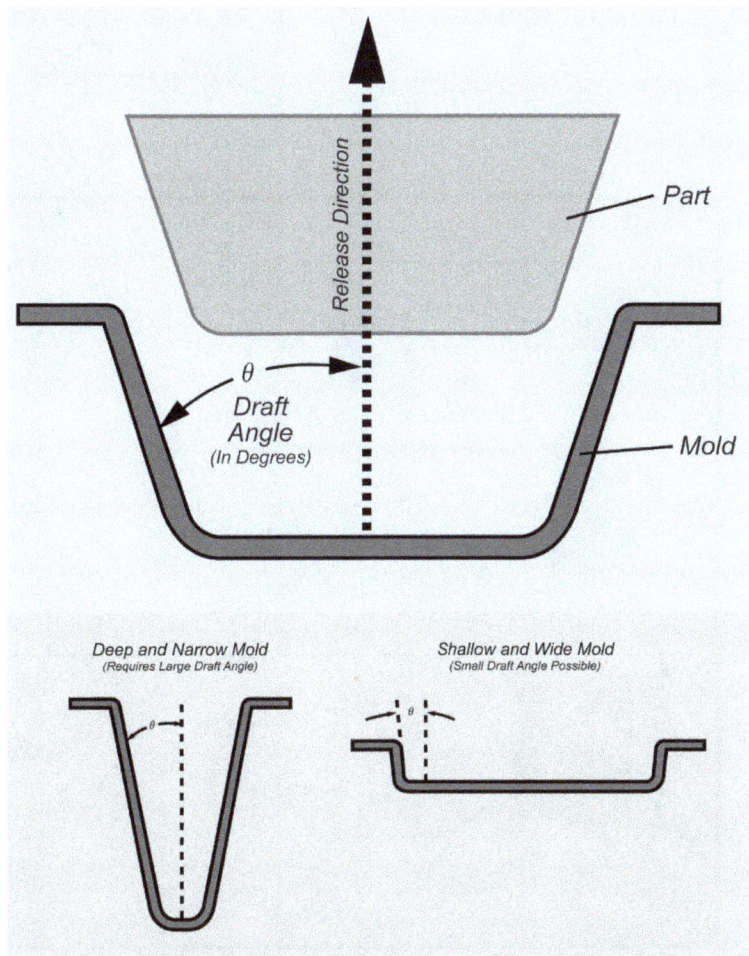

Draft is the angle of a mold's walls in relation to the direction of a part's release from it.

9

feature that would otherwise make it difficult to remove from a mold (such as "negative draft" or "undercuts") alternate molding systems can be developed for this purpose. Such mold systems may include "multi-sectioned" molds or ones that facilitate removable "inserts". A multi-sectioned mold is one that allows the mold to break apart into multiple pieces so the component can be easily extracted from the mold. The flat flange surfaces where these mold sections meet is referred to as the "parting plane" because this is where the mold "parts", or separates. The line on the mold faces, where this parting plane comes in contact with the actual molded part, is called the "parting line".

For parts with tight details or corners, "inserts" can be used to hold the laminate against the mold until the composite cures. However, these types of mold systems are not usually required except for the most difficult or demanding situations. Skilled and wise composites fabricators aim to keep their parts as simple as they can realistically manage, paring them back as needed to minimize any potential molding headaches—so, likewise, design your parts with simplicity in mind.

When it comes to mold-making, the actual materials used to fabricate a mold are only limited by a few factors. The material used in making a mold should at least be able to:

• Create the proper part shape
• Accommodate the chemicals used in the molding process
• Produce smooth, low porosity mold surfaces
• Maintain its shape while resisting the rigors of lamination and demolding

SHAPEABILITY

A properly shaped mold is imperative to a properly shaped part. Some materials lend themselves to being shaped much more readily than others while some materials require some very specialized skills to be shaped well. Indeed, the fabricator's own familiarity and skill with certain materials may ultimately dictate which mold-making materials they choose to use. Sheet metal can be bent, formed, cut, welded, and finished to create various mold shapes, but complex surface detail and contouring may require some significant skill to

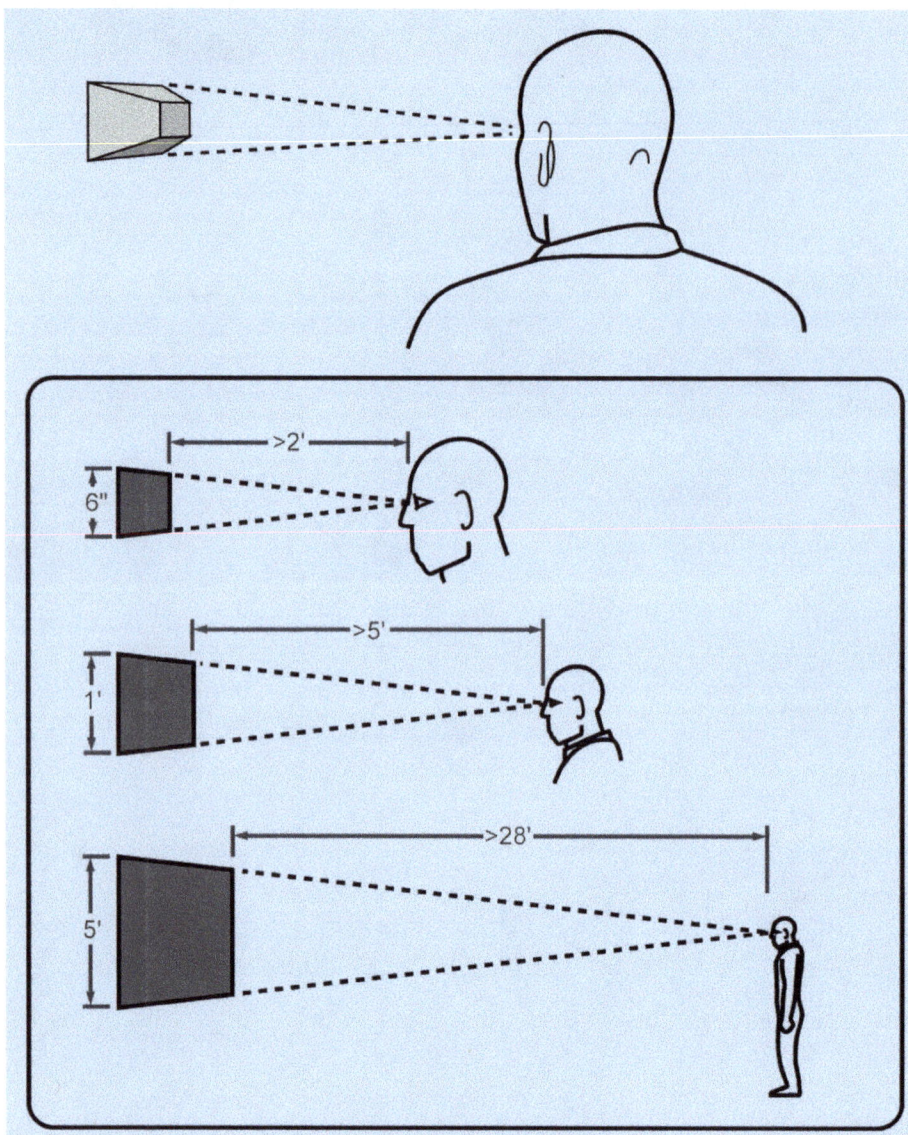

With proper draft, all molded sides should be visible when spaced at an adequate distance. (Example distances are shown for parts with 5 degrees of draft.)

10

produce. Wood will shape relatively easily with common woodshop tools, but it may need to be glued-up from multiple boards or cut from large plywood stock to develop the right size for many molded projects. Large, compound-curved surfaces can be especially troublesome to fabricate out of wood without some serious planning and craftsmanship (as found in classic wood boat building). Plastic sheet can be very helpful in creating both planar and compound curved surfaces, if you're familiar with using it. Large forms can be easily carved and sculpted out of foam (whether urethane, styrene, or other types), especially to make shapes that are too difficult or costly to form in other traditional materials, although foam has its own cost and practical size limits as well. Plaster and clay are favorites for sculptors, and can shape very well, but can be expensive for large shapes and are generally fragile to use.

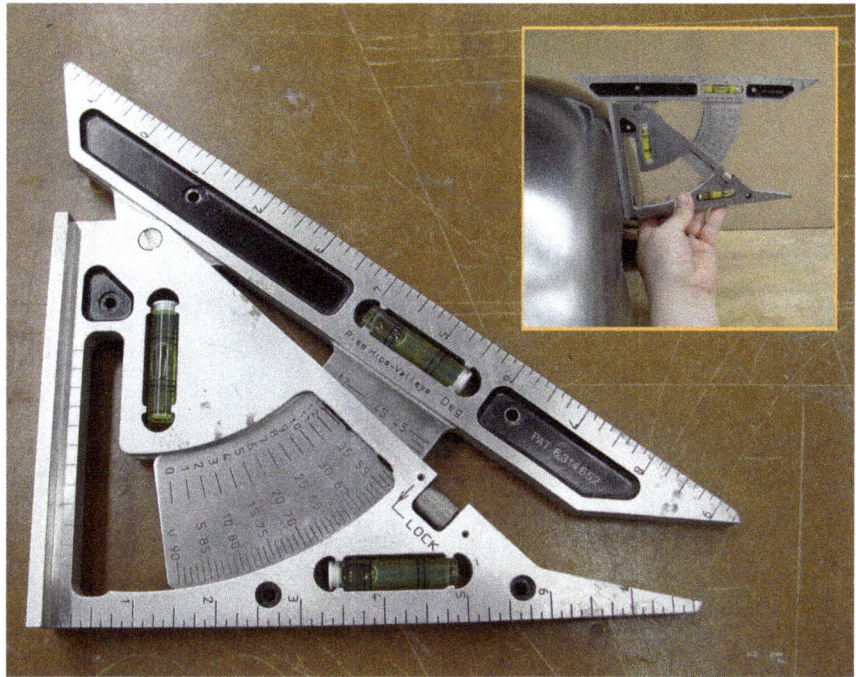

A protractor will work well for measuring draft angles, but a "pivot square" (shown above), available from the C.H. Hanson Company (www.chhanson.com), can be especially helpful.

CHEMICAL COMPATIBILITY

Compatibility between a mold and the resin system formed in it can be critical to maintaining a good composite molding system. Avoid resins or other molding chemicals that may attack the mold. This is especially true of polyester and vinylester resin systems that are to be used in conjunction with styrene-based plastics or foams. The styrene monomer in the resin can begin to attack and dissolve the mold. Further, a silicone mold used in contact with polyester resins will inhibit the polyester from curing, even when the resin is highly catalyzed. However, such chemical interactions are rarely a problem with metal-surfaced molds.

SURFACE SMOOTHNESS

A mold's surface should be as smooth as possible for two reasons: to aid in the demolding of the cured part, and to

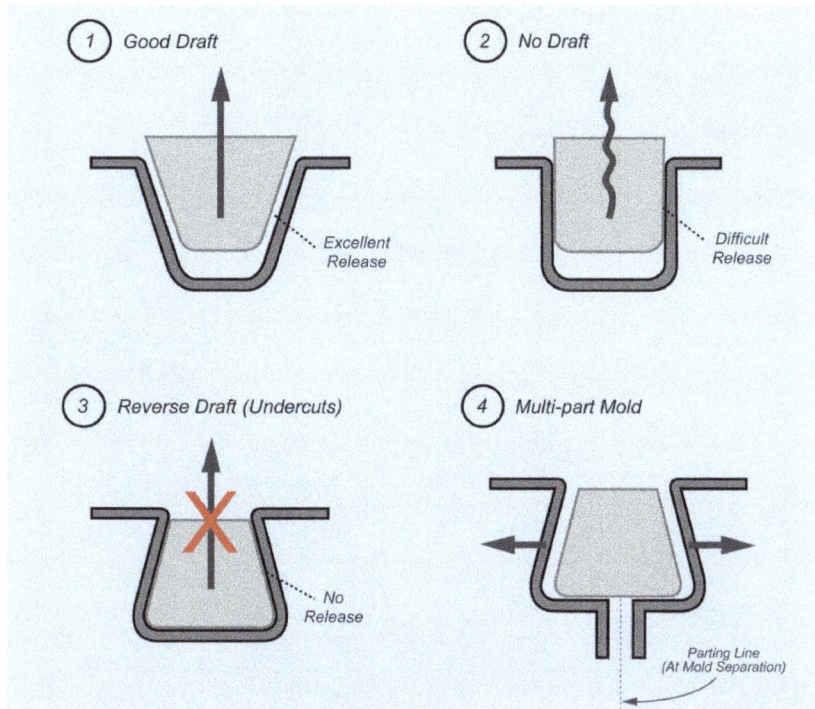

A properly drafted part will release easily from a mold. With undercuts, multiple mold sections are necessary.

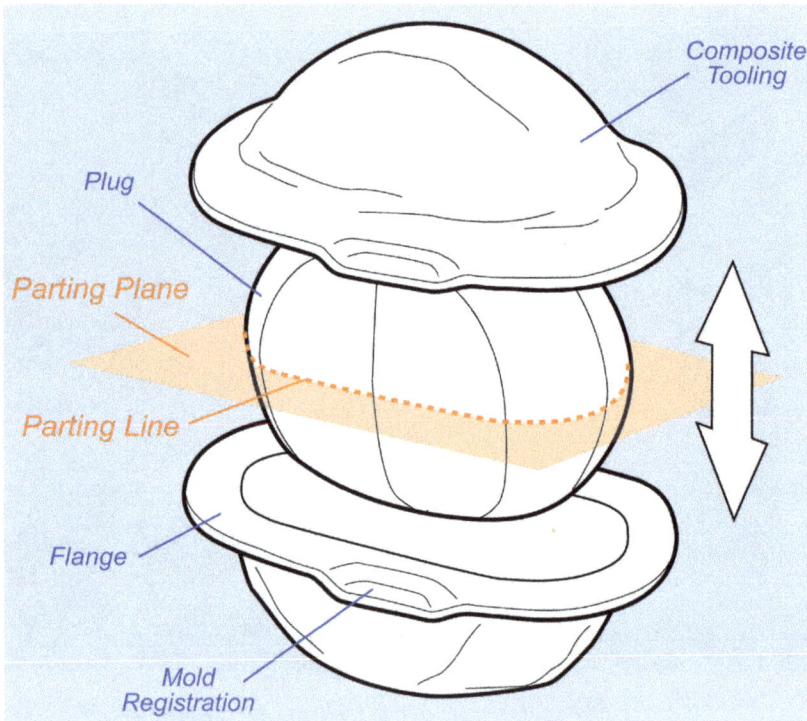

Illustration of the parting plane, parting line, flanges, and mold registration found on a two-part mold.

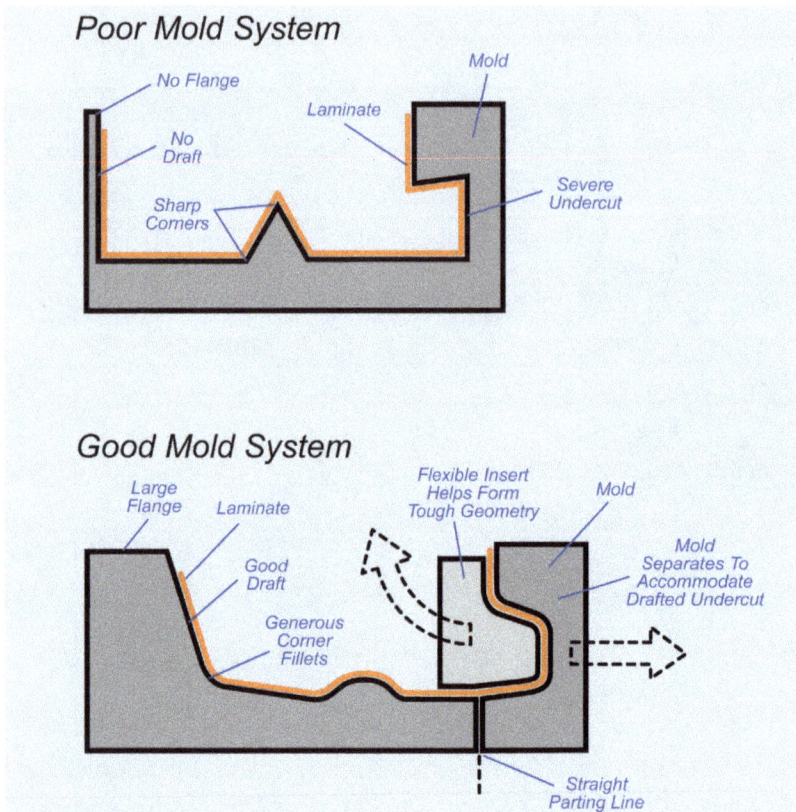

Examples of good and bad mold features.

minimize additional finishing of any parts made in the mold. The thermoset resins used in composites will very accurately reproduce texture and detail from mold surfaces down to a microscopic level. However, texture and detail (especially pits and pores) will create extra "bite" for the resin to lock into, making release of a composite part much more difficult or impossible. Unless they are absolutely needed, details should be minimized, and textured surfaces should be replaced with mirror smooth surfaces as much as possible so parts will easily release from the mold. Further, unwanted texture on mold surfaces will transfer to every part formed in the mold, requiring considerable work on all molded pieces to remove them. To avoid these surface finishing hassles, spend the time necessary to ensure a smooth, high quality surface. This will include repairing any imperfections and polishing the mold surfaces to a high gloss using cutting/rubbing compounds (as shown in Handbook - 1). Polishes can be applied by hand (for small jobs), or with the assistance of air or electric-powered buffing equipment. At the very least, all mold surfaces in contact with a curing laminate should be highly reflective and free of visible scratches—just as a skillfully polished automobile paint job would be.

STRENGTH

In order to provide a long, successful service life, a mold must be strong enough to handle the rigors of numerous lamination, curing, and demolding cycles. To ensure sufficient longevity for your mold, build molds that are reinforced adequately enough to not deform under the weight and size of the parts you will be forming in them. Always "overbuild" a mold so it will be at least 3 to 4 times the strength of the composite formed in it. To further minimize mold flexure, include reinforcement ribs as needed by either bonding or mechanically fastening them to the outside of the mold. In cases where high pressure lamination methods will be used (such as with compression molding or expandable inserts), correspondingly reinforce the mold or build it with robust enough materials to ensure its integrity throughout the molding process.

MOLD-MAKING MATERIALS

Several mold-making materials will be demonstrated in this book, but there really are no limits to the types of materials that can be used to build a mold, as long as they follow the general criteria listed above. In fact, one of the most intriguing aspects of mold-making is the practice of solving a molding challenge using a creative mix of materials and molding methods. Indeed, much of the expertise needed to become a competent mold-maker comes through hands-on experience, so several varied techniques will be covered in this book.

Lastly, because mold-making materials come in a wide range of cost, shapeability, and durability, each material has a molding application for which it is generally best suited. Because of how well they meet common mold-making criteria, metals (in sheet, billet, or cast form) tend to provide the best strength and overall longevity for mass-production composites, although they can require costly and time-consuming steps in their fabri-

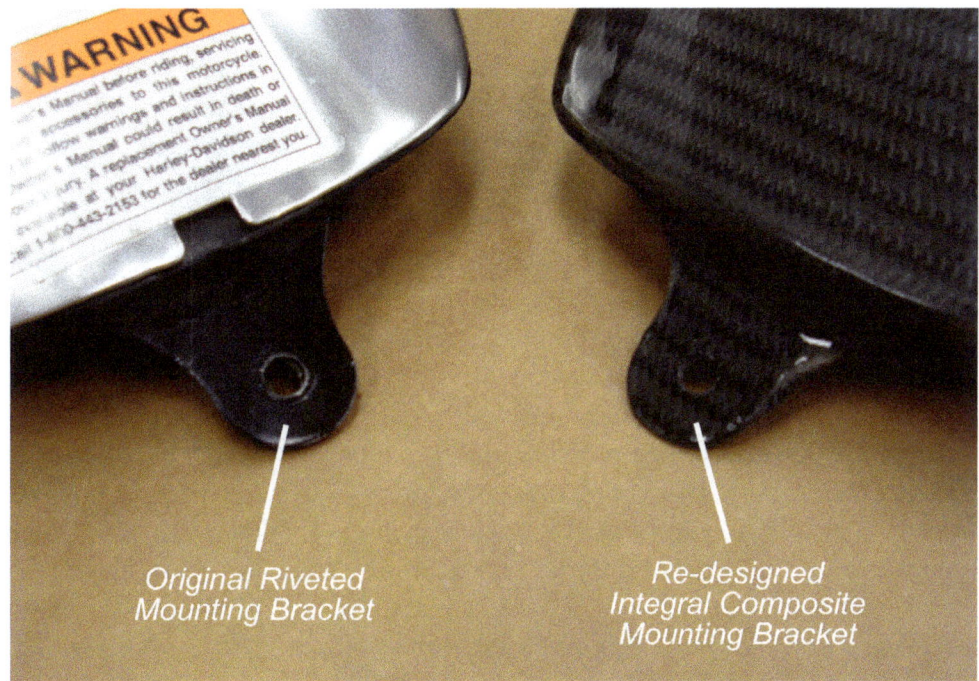

Some composite parts can be designed to simplify their metal counterparts, as shown.

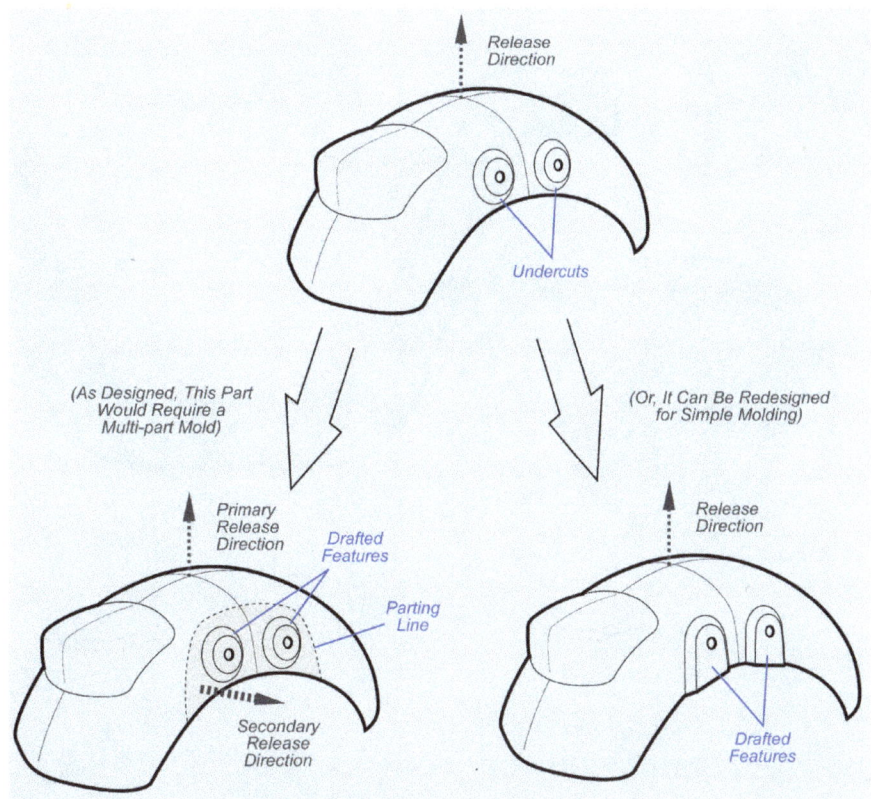

The geometry of the part will largely dictate how the molds should be shaped. Even slight geometry changes can often drastically simplify the mold system, as shown.

cation (as shown in Chapter 5). Woods are a good, low-cost option, especially as reinforcements in a mold system, but they must be sealed with a coating of resin or other non-porous material wherever they will be in contact with the laminate (as shown in Chapter 6). Thermoplastic sheet works well in molds since it is generally non-porous to begin with and can be readily formed using a heat gun or oven. When amorphous thermoplastics are used, they can usually be reinforced with composites (as shown in Composites Fabrication Handbook - 1), and can be polished to a very high gloss surface. Cast thermoset plastics (as demonstrated in chapter 6) can be used to form very robust, non-porous molds, and can be machined to fit exact tolerances. Composite materials are also very helpful for building composite molds (called "composite tooling"), and can be very cost-effective and durable for most medium to large mold applications (as shown in chapter 2). Even various types of foam (both rigid and expanding types) are useful for different mold-making tasks (also shown in chapter 2) and for roughing out complex, curved shapes. Lastly, plaster or clay can be used to produce very intricate and ornate shapes, though both of these materials are somewhat limited in their use as molds because they are very fragile and their molding surfaces must be specially sealed. (See the chart at the end of this chapter for a general overview of several different mold-making materials).

CHAPTER CONCLUSION

When developing a new mold for a composite laminate, carefully consider its shape and what materials would be appropriate to create a mold for that shape. Determine which surfaces would be best formed in direct contact with the mold faces, and develop a mold that is smooth enough for the parts molded in it and strong enough to handle the lamination and demolding processes. Likewise, be sure to select mold-making materials that will be chemically compatible with the laminate in order to ensure good usability and overall longevity of the mold. As each possible mold-making material has applications for which it is best suited, study out the material options and select ones that will work most effectively for your particular molding needs.

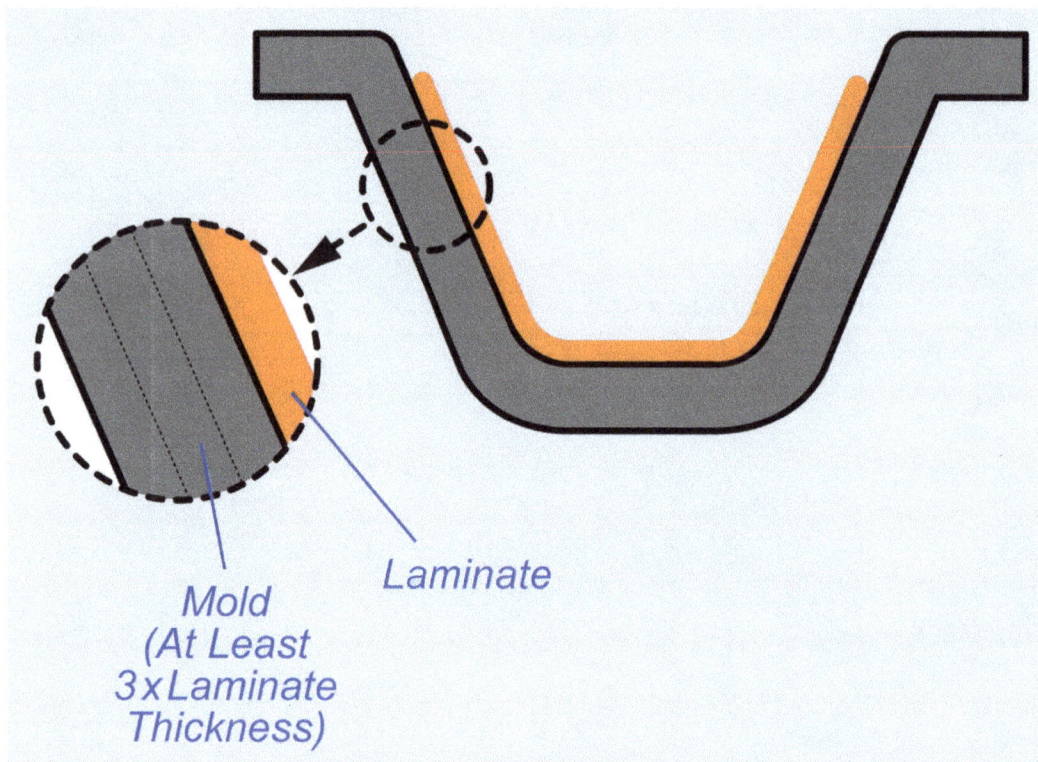

Composite tooling should be three to four times the thickness (or strength) of the composite formed in it.

Laminate

Mold
(At Least
3 x Laminate
Thickness)

	Recommended Mold Shapes	Cost	Mold-forming Tools	Mold-making Speed	Mold Toughness and Longevity	Surface Finishing Guidelines	Best Used For...
Wood	Best For Planar & Simple Curved Surfaces (Compound Curves Are Difficult to Produce)	Low to Moderate	Common Woodworking Tools	Moderate To Fast	Low To Moderate (Dependent On Density of Wood Or Use of Ply)	Sanding, Resin/Lacquer Sealing, And Polishing	Large Molds With Very Low Production Volumes. Excellent For Reinforcing Molds Or For Plug Building
Sheet Metal	Best For Planar & Simple Curved Surfaces (Compound Curves Are Difficult to Produce)	Low to Moderate	Common Metalworking Tools	Slow To Fast	Moderate To High (Dependent On Sheet Gauge and Geometry)	Filing, Sanding, And Polishing	Small To Moderate Size Molds. Excellent For All Resin Types
Bulk Metal (Cast or Billet)	Planar and Compound Curved Surfaces Possible	Moderate to High	Metal Machining Tools	Slow To Moderate	High	Filing, Sanding, And Polishing	Small To Moderate Size Molds With High Production Volumes and/or Forming Pressures And All Resin Types
Thermoplastic Sheet	Planar and Compound Curved Surfaces Possible	Low to Moderate	Common Woodworking Tools and Heat Gun Or Oven	Moderate To Fast	Moderate (Dependent On Sheet Thickness and Plastic Type. Some Sensitivity To Impact)	Sanding, And Polishing	Small to Moderate Size Molds. Fast, Inexpensive Mold-making For Low Production Volumes
Bulk Thermoset (Cast or Filled)	Planar and Compound Curved Surfaces Possible	Moderate to High	Metal Maching Tools	Slow To Moderate	Moderate To High	Sanding, And Polishing	Slightly Lower-cost Alternative To Bulk Metal Molds For SmallTo Moderate Size Molds.
Composite (F.R.P.)	Planar and Compound Curved Surfaces Possible	Low To Moderate	Common Woodworking Tools	Moderate	Moderate (Dependent On Wall Thickness and High-Temp Use)	Sanding, And Polishing	Relatively Fast Mold Making At Low Cost. Small To Large Size Molds. Lightweight and Durable Molds Possible With Moderate Production Volumes
Foam	Planar and Compound Curved Surfaces Possible (Low Detail Is Best)	Low To High	Common Woodworking Tools	Moderate To Fast	Low To Moderate (Low-density Foams Are Easily Damaged)	Sanding, Filler Or Resin Sealing, And Polishing	Construction Of Plugs, One-off Forms, Mold Cores, Or Very Low Production Volume Molds
Plaster	Planar and Compound Curved Surfaces Possible (Low Detail Is Best)	Low	Common Woodworking Tools	Moderate To Fast	Low (Very Brittle And Easily Damaged)	Sanding, Resin/Lacquer Sealing, And Polishing	Construction Of Plugs, One-off Forms, Or Mold-cores
Modeling Clay	Planar and Compound Curved Surfaces Possible (Low Detail Is Best)	Moderate To High	Clay Sculpting Tools	Slow To Moderate	Low (Soft And Easily Damaged Surfaces)	Smoothing and Multi-step Sealing	Small To Moderate Size Plugs Or As A Filler For Mold Gaps And Corners

A comparison chart of various mold-making materials.

Chapter Two

Composite Tooling Fabrication Techniques

Building Durable Molds Out of Composites

CHAPTER INTRODUCTION

Molds that are fabricated from composite materials are referred to as "composite tooling". Composite tooling is relatively easy to create and produces high-quality parts suitable for a wide variety of applications. Because of how beneficial composite tooling is in mold-making for composites, this chapter describes the process in depth.

GUIDELINES FOR FABRICATING COMPOSITE TOOLING

The term "composite tooling" comes from the plastics molding industry where molds are referred to as "tools". The actual process of cre-

Composite tooling is cost-effective, durable, and lightweight - perfect for many composite molding needs.

ating composite tooling was developed decades ago, likely as a realization that composites worked very well at creating molds for other composite parts. Composite tooling is extremely cost effective for molding exceptional high-quality parts in low to moderate quantities, and can be used in conjunction with several different molding procedures.

Fabricators who are familiar with typical wet layup methods generally find that building composite tooling is relatively straightforward. In fact, apart from a few extra steps, composite tooling fabrication is exactly the same as the average wet layup procedure. These fabrication steps include:

- Creating and sealing the mold shape (plug construction)
- Attaching flange forms
- Prepping for layup
- Applying tooling gel coat or surface coat
- Laying up the mold
- Reinforcing the mold (if necessary)
- Curing the mold
- Demolding, trimming, and finishing the new mold

STEPS TO PLUG CONSTRUCTION

The first thing to consider in making composite tooling is that the mold itself needs to be formed over another mold, plug, or existing part. While several other types of mold-making materials are self-supporting, uncured laminates do not possess any significant rigidity of their own and therefore require extra steps in their creation to establish a form over which they can be fabricated. These extra steps, though, give the fabricator the benefit of assessing the final part's shape and size long before making the mold and demolding that first completed laminate. If you first make a plug that represents the actual finished shape and dimensions of your desired component, a composite mold can then be created over this plug and it will replicate your component exactly. As an analogy, the old junior high school trick of creating a "wax finger" demonstrates this very well. After dunking and

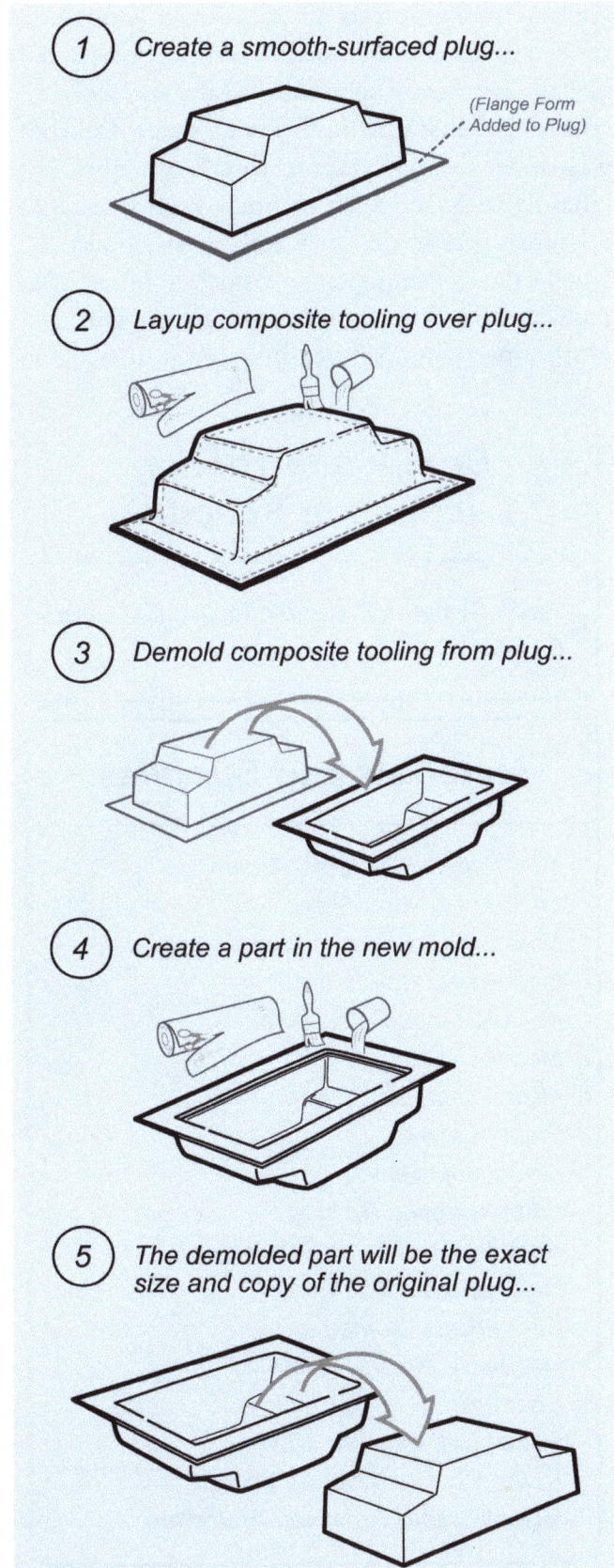

1. Create a smooth-surfaced plug...

(Flange Form Added to Plug)

2. Layup composite tooling over plug...

3. Demold composite tooling from plug...

4. Create a part in the new mold...

5. The demolded part will be the exact size and copy of the original plug...

The steps to create a copy of a plug or part using composite tooling.

cooling the liquid wax on your finger enough times to create a thick shell, carefully cutting it in half and removing it will reveal a mold of your finger that exactly copies its shape - all the way down to fine details in your fingerprint. For circumstances in which you may want to make a composite version of an already existing part (such as a car manufacturer's stock hood or body panels), it is relatively simple to use the part as a kind of pre-manufactured plug to create a

The metal fender blank, polished, waxed, and ready for use as a plug.

Composite Tooling Lamination Schedule

Resin System: Polyester Tooling Gel Coat
6 Layers - 1.5 oz. Fiberglass Mat

Materials and Supplies

10 - 15 lb Polyurethane Foam
MDF or Plywood Sheet (3/4" Thick)
Non-porous Sheeting (plastic, or metal)
Corrugated Cardboard
Mold Release/Parting Wax
Modeling Clay
Mixing Cups and Sticks
Paintbrushes
Disposable Gloves
Foam Shaping Rasp
Sanding Block (with Sandpaper)
Ruler and Marking Pen
Clean Rags/Cloths
Naptha or Mineral Spirits
Scissors
Power Drill with Drill Bits
Rotary Tool with Cut-off Wheels
Miscellaneous Fasteners/Hardware

The composite tooling lamination schedule, along with materials and supplies, used in this demonstration.

dimensionally correct mold. The extra work required to build a proper plug or to copy an actual part will definitely pay off when you see just how well that first finished component comes out of the mold.

To create a plug from which to form composite tooling, first plan to address the previously mentioned issues of shape simplification, draft, release direction, flange location, and parting plane location (if applicable). Once these primary concerns are resolved, determine the appropriate materials necessary to build a satisfactorily robust, smooth-surfaced plug. At every stage of plug construction, use good craftsmanship and proven building methods - the quality of your mold (and every part made from it) depends on conscientious fabrication of the plug. After completing construction of the plug, seal it, and prepare it for layup with appropriate release agents.

As an exercise in plug construction and corresponding composite tooling fabrication, we will demonstrate building a plug and a mold for a motorcycle rear fender that has a specially designed rear stop light and mounting points.

This demonstration will use a combination of different fabrication methods, including employing an existing part and shape templates, forming flanges, and making a multi-section mold. Using an existing steel fender blank for this demonstration will save considerable time in making the plug because it already has the correct shape and smoothness needed (especially after polishing), as well as the rigidity to support the weight of all the mold-making materials that will be placed on it.

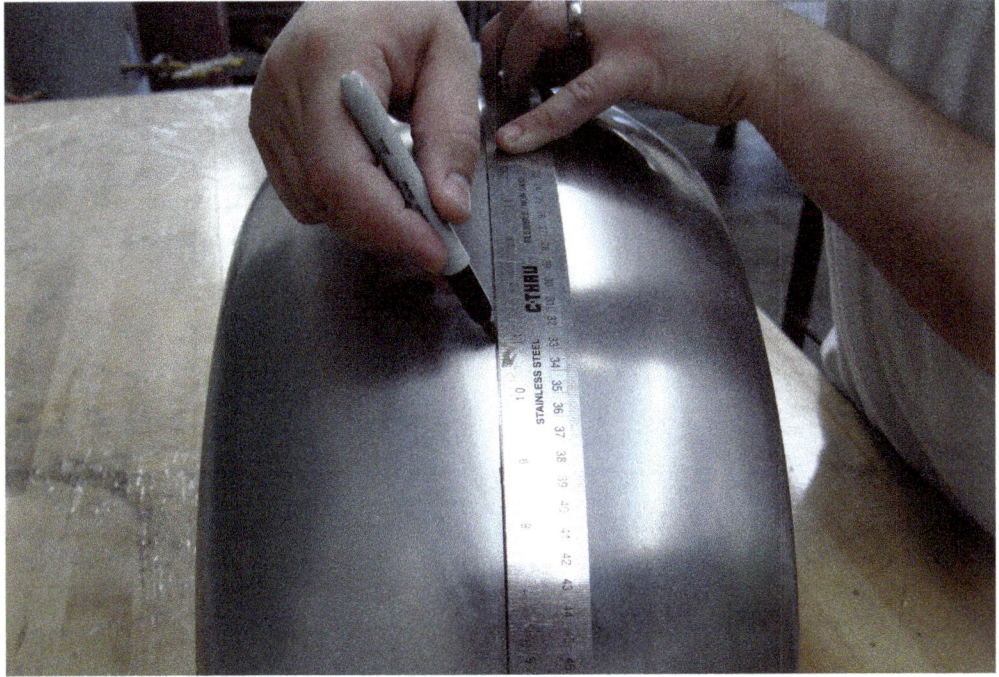

Additional design features can be easily adhered to the steel fender blank, including shaped foam, metal, wood, or plastic details. For example, the protruding mounting points on the side of the fender will be created from shaped plastic bolted to the side of the plug, and a stop light will be constructed from foam and adhered to the top of the plug. Since the mounting points will create undercuts on the mold, the mold will be created in two sections to accommodate these details.

Mark the center line of the plug with a square and permanent marker to help register any features that will be added

Continued on page 22

Use a flexible ruler to mark the centerline over the length of the plug.

Use a template to trace the outline of the brake light feature, along with its station locations. Make sure the feature outline is centered correctly on the plug.

19

Use templates to trace the shapes of the station forms onto urethane foam sheet. Cut out the station forms with a bandsaw.

Mix some quick set epoxy superglue or wood glue will work as well) and apply it between foam sections. Avoid spreading any epoxy near the seams as this will form waves when the foam is sanded.

Use a rasp to shape the bottom of each foam section...

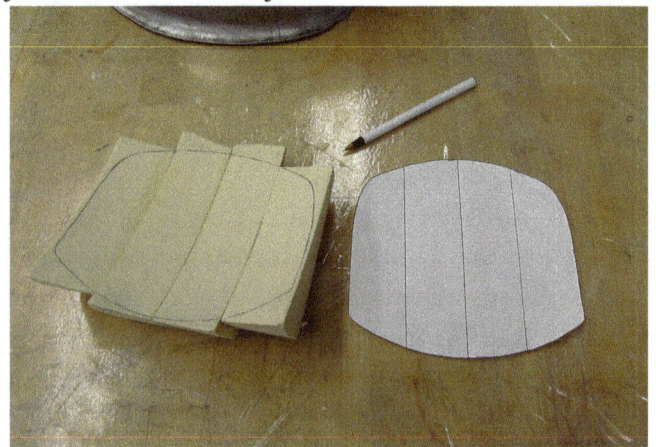

Trace the template on the bottom of the sections...

...until it fits the fender's contour very well.

...and begin shaping the foam using a sanding block.

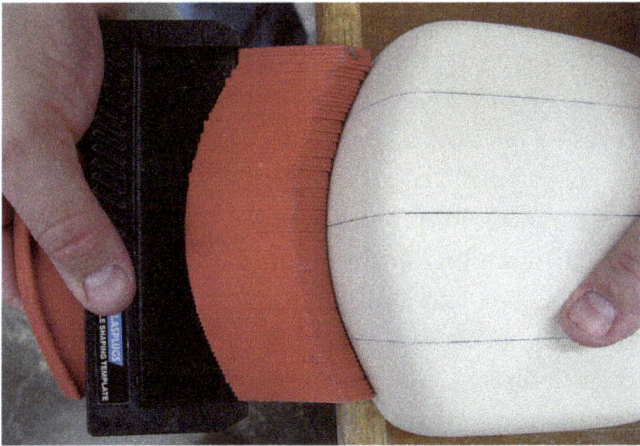

Press a shaper gauge onto the surface…

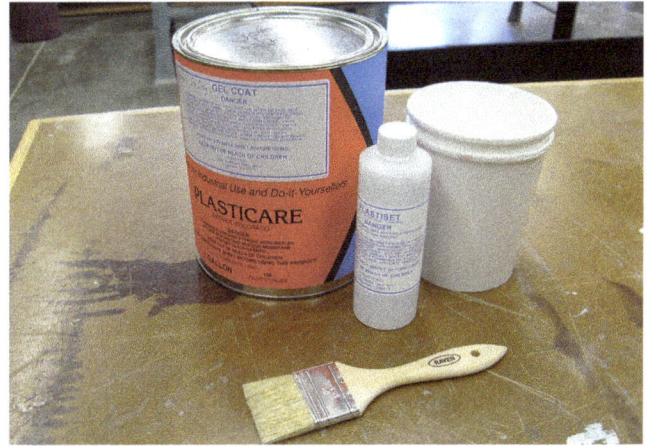

Polyester tooling gel coat will be used throughout this project to first seal the foam, and then later to create the actual composite tooling.

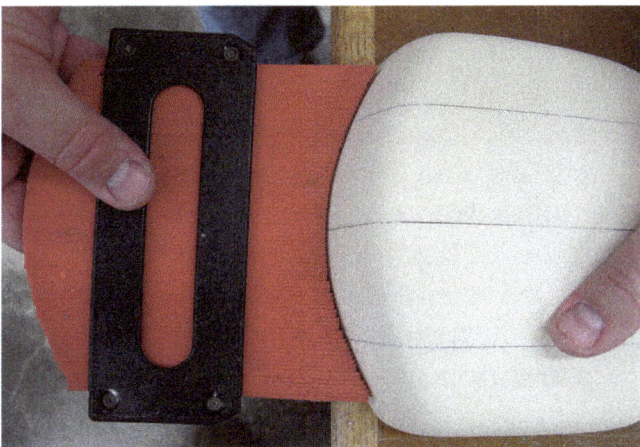

…and then flip it over and check it to see if the surfaces are symmetrical.

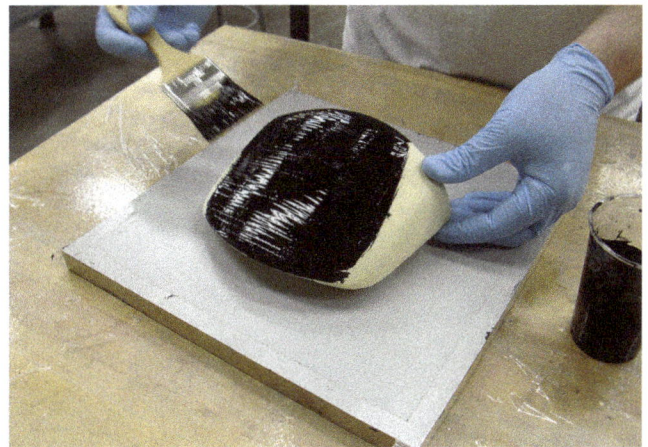

Mix the tooling gel coat as recommended by the manufacturer and apply it onto the surface of the foam with a brush.

Once completely sanded, use an air hose to blow out any dust caught in the foam's porous surfaces.

Continue applying the tooling gel coat until the foam is completely covered. Once cured to a tacky state, apply a second coat to produce a thick, sandable surface.

Wet-sand the surface with progressively finer grits until it is smooth...

...and then use cutting/rubbing compound to further polish the surface...

...followed by three to four coats of parting wax.

to it. Include any other reference lines that may be necessary to help position other part details. Templates can be helpful to properly place features in relation to the plug's reference lines.

Use good fabrication practices to construct a high quality plug. For example, prior to constructing a plug, ensure that it is being built upon a level and straight surface. When shaping the plug, use templates, guides, and accurate measurements. When constructing shapes with templates set apart at station points, provide proper, measured spacing between these station points to develop the shape correctly. Simple shapes may only need a couple guide sections to adequately describe the form whereas complex shapes may need several guide sections. If using guide sections, ensure that each one is mounted vertically and rigidly during construction.

For this demonstration part, shape templates were traced onto polyurethane foam, cut out using a bandsaw, shaped to fit onto the plug surface, bonded together, and then sanded to match the template lines. Polyurethane foam is ideal for the rear stoplight detail on this plug because it shapes quickly, will seal to a very smooth surface with resin, and has enough compressive strength to handle the weight of the composite tooling to be laid-up over it.

Create a hard, smooth coating over any porous plug surfaces using resin (including epoxy surface coat, or polyester tooling resin, resin-based primer, lacquer, urethane, or a coated composite laminate. Try to apply any surface coatings as smoothly as the particular coating liquid will allow because any imperfections will need to be sanded and smoothed completely. Build this coating layer as thick as needed (through multiple applications, if necessary) to ensure that it will not be sanded through very easily while smoothing out any surface flaws.

Bring the final sealed surface as close to mirror-quality as possible by wet-sanding it to a high gloss (of at least 600 grit - though higher grits are recommended), followed by additional smoothing with cutting/rubbing compound and high quality polish. Use good craftsmanship so the final surfaces will be smooth, free from waves, bumps, and pits so they will maintain their correct overall shape. Wax and buff the plug with release wax, applying a minimum of four coats. As shown in this demo, coated urethane foam components can be attached to the main plug with double-stick tape, followed by seam filling with modeling clay.

FLANGE FORM CONSTRUCTION

"Mold flanges" are wide, flat sections that extend from the edge of a mold's surface. These flanges are used for a few reasons which include: helping to make a mold more rigid and manageable (especially during handling and throughout layup procedures), "locating" or "registering" multiple mold sections together, or acting as a surface onto which sealing tapes can be adhered for advanced vacuum bagging procedures. While some fabricators may be tempted to forego the construction of mold flanges, flanges can be extremely beneficial if for no other reason than to strengthen the lip of the mold to aid in both mold longevity and to ease part demolding. When creating composite tooling, it is common to incorporate mold flanges into the mold at the time of the mold's layup, so designate their location and shape before you begin the layup. The steps for creating these flanges in composite tooling are described and demonstrated in this chapter.

Plan on how to accommodate flanges for the plug, and where the parting plane and flanges should be placed early on in the fabrication process. If you intend to create vacuum-bagged parts in the mold (as shown later in this book), plan to make the flanges at least 2 to 4 inches wide, otherwise, 1 to 2 inches of flange width should suffice. Flanges for vacuum-bagging molds need to be as smooth as possible to avoid vacuum leakage from poor sealing. However, the flanges used in multi-section molds may require additional width, alignment hardware, or regis-

Use double-sided tape to adhere the brake light feature.

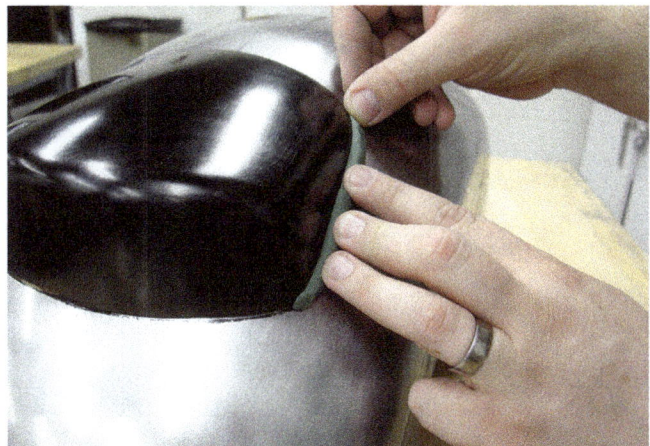

Apply a bead of clay around the seam between the brake light and steel plug using finger pressure to smooth it into a nice fillet.

Further smooth the clay with a round-ended stick.

From plastic, shape the fender mounting features...

...and bolt them onto the plug.

tration details (as demonstrated in this chapter) to make sure the mold sections fit together correctly.

The flanges on composite tooling are most easily (and durably) made by laying them up at the same time as the rest of the mold. To provide a support for the mold's flanges during layup, firmly attach a set of shelf-like forms at the parting plane. The most appropriate method for making these temporary flange forms depends upon the particular type, shape, or size of plug; small plugs can use lightweight materials such as cardboard or foam core as flange form supports, whereas large molds may require more elaborate wood or metal constructions.

Flange forms are most easily created when they are attached directly to the bottom edge of a plug, simply because a flat surface under the plug (such as a table top) can act as an effective flange form. Odd-shaped plugs may require more involved attachment methods for the flange forms. Bent or formed sheet materials are especially helpful for such shapes.

Some materials that make good flange forms include plastic sheet (such as acrylic, styrene, polypropylene, or high-density polyethylene cut with woodworking tools and molded as needed with a heat gun), sheet metal (steel or aluminum sheet stock cut and formed with metalworking tools), or wood sheet (melamine/Formica faced board, MDF, or Masonite cut with woodworking tools). Again, note that porous materials, like wood or foam, will require complete sealing with resin, lacquer, paint, or primer to a smooth surface before proceeding.

Flange forms can be secured to the support structure using modeling clay or adhesive-bonded ribs cut from sheet material. Mount these flange forms at least an inch below the intended trimline of the final part. This will provide enough space for "flash" (or extra, unusable edge material) to be trimmed away to make a clean finished edge on the parts formed in the mold.

Keep in mind that the flanges on a composite mold form right-angled corners where they meet the plug - a shape that is atypical of most other composite geometries. Although it is usually best to avoid sharp angles and tight corners, right angles at the flanges of composite tooling are common, but as they are created, they are laid up in a way that minimizes voids while sufficiently reinforcing them, as will be demonstrated.

GENERAL LAYUP PREP

Decide on what reinforcements and resins to use in building the mold as well as what materials you will actually layup into the mold. As you determine the

Continued on page 28

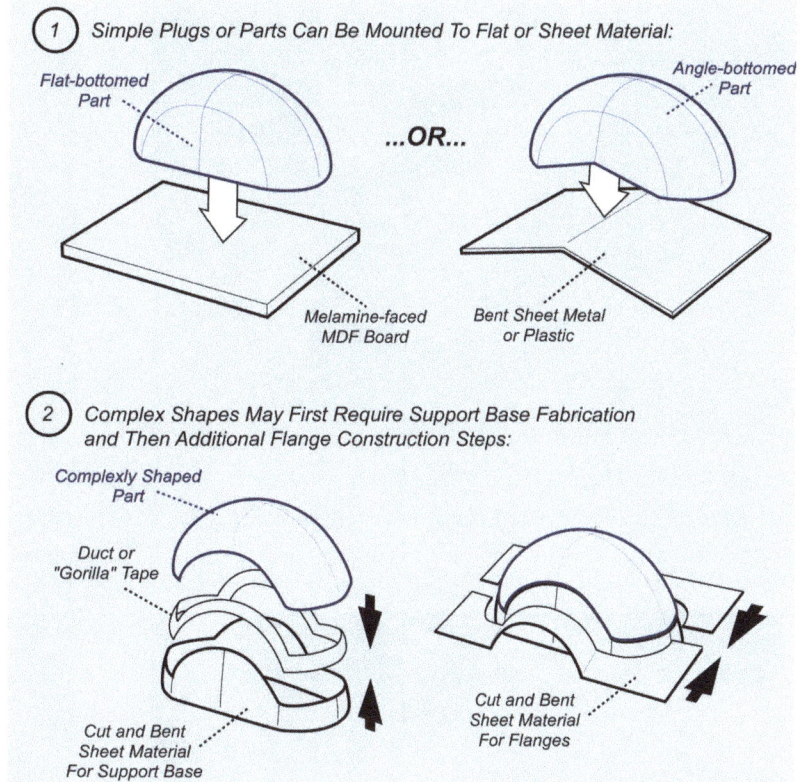

1. Simple Plugs or Parts Can Be Mounted To Flat or Sheet Material:

Flat-bottomed Part

Angle-bottomed Part

...OR...

Melamine-faced MDF Board

Bent Sheet Metal or Plastic

2. Complex Shapes May First Require Support Base Fabrication and Then Additional Flange Construction Steps:

Complexly Shaped Part

Duct or "Gorilla" Tape

Cut and Bent Sheet Material For Support Base

Cut and Bent Sheet Material For Flanges

Several methods can be used to create flanges.

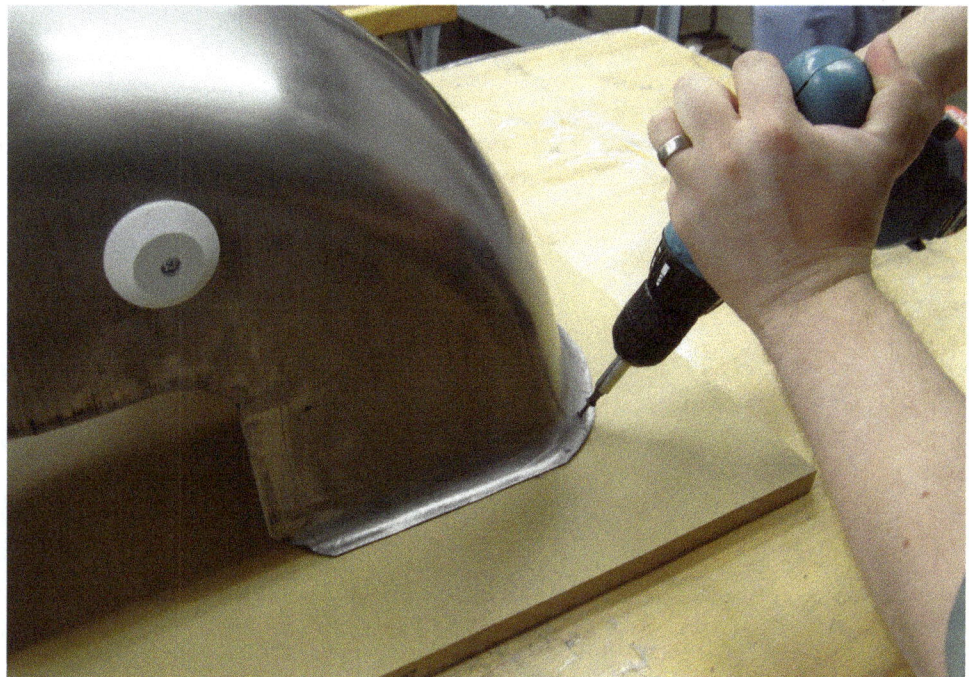

Secure the plug to a base...

25

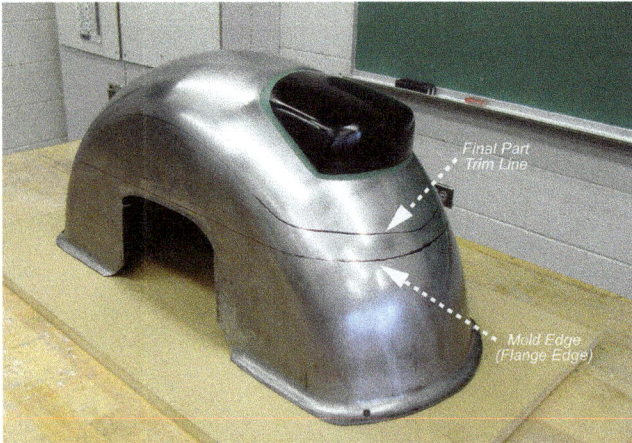

…and mark the location of the final part's trim line as well as the location of the mold edge and flanges.

Use this template to create multiple flange form support sections using cardboard. (An "eggcrating" method is shown here.)

Mark and trim cardboard to match the shaping of the fender where the mold edge is positioned…

Fold cardboard to make additional flange form supports for the sides of the plug and attach them with double-sided tape. Use additional masking tape as needed.

…and then trim another template to match the vertical curvature of the fender.

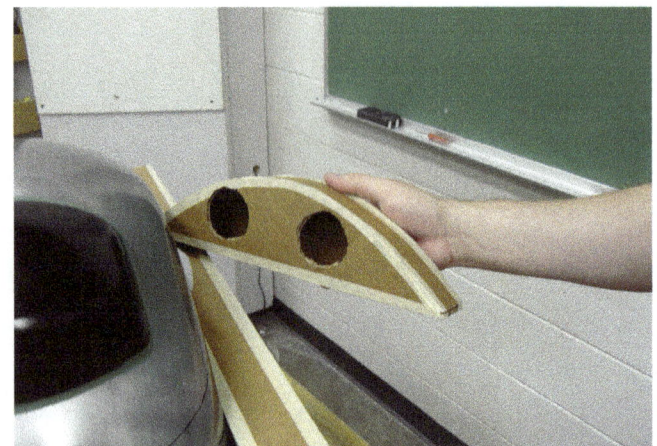

Create a support that will help form the flanges for the removable mold section…

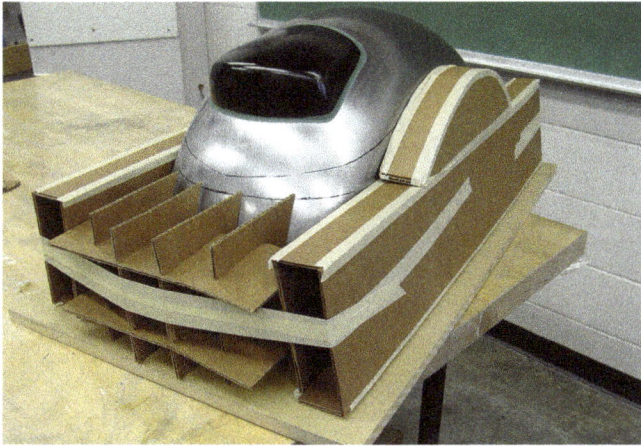

...and attach it to the plug as well.

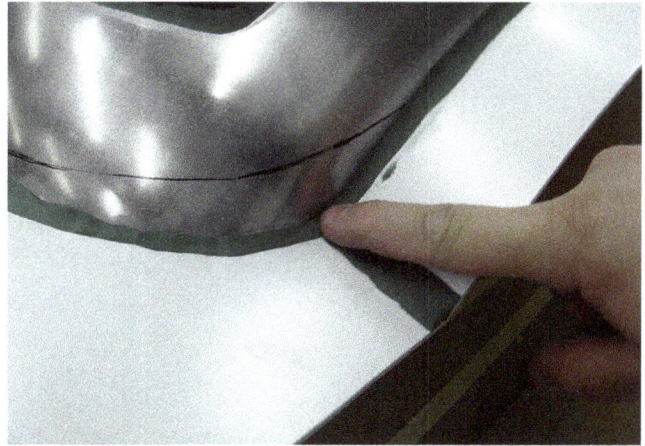

...and then use clay to fill any gaps between the flange forms and plug.

Using the previous cardboard template, cut smooth plastic to size to match the flange form supports and contours of the plug. Apply these flange forms to the supports using double-sided tape.

Use a piece of plastic to square off the clay only at the areas that will be part of the removable mold section.

Wax all surfaces of the flange forms and plug...

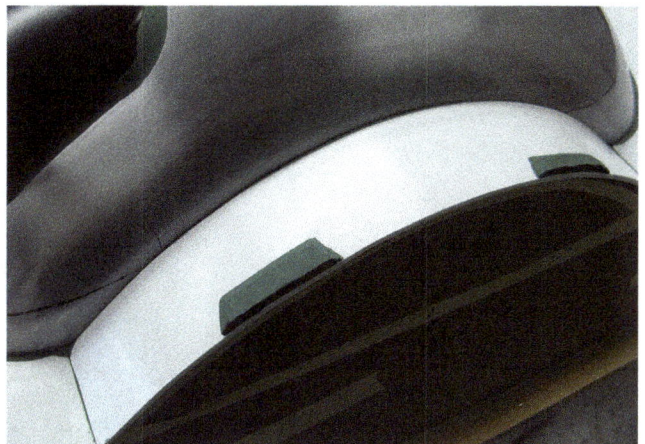

Create small clay details to form mold registrations in the flanges on the removable mold section.

Mix the polyester tooling gel and apply it to all the mold and flange surfaces.

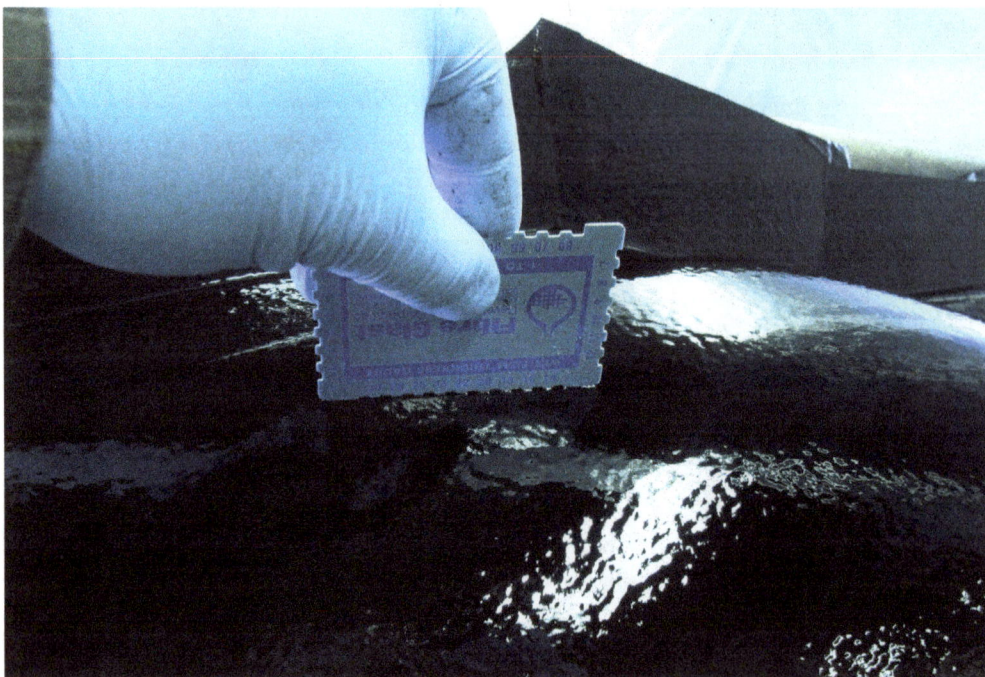

A gel coat gauge pressed into the gel coat will show the resin thickness by the longest depth "finger" that is completely covered.

actual matrix and reinforcements for your mold, bear in mind that the mold itself can be constructed out of polyester or vinylester resin and fiberglass at relatively low cost - even if you intend to form expensive room-temperature epoxies and carbon fiber laminates in your mold. In fabricating composite tooling, inexpensive resins and reinforcements can often work just as well as pricey ones, as long as the added performance and properties of more costly materials is not really necessary.

Have enough resin and reinforcement on hand to complete the mold before starting the layup by approximating your material needs beforehand using simple measurements and weights. Weight ratings for fabrics are provided wherever they are sold, but beware that fiberglass weights can differ from all others since it is often designated in weight per-square-foot, whereas other fabrics are represented in weight per-square-yard. Measure the overall length and width of the mold in either yards or feet (depending on whether you are using fiberglass or other fabrics), rounding up to the next highest number if the measurement is a fraction, and then multiply the two measurements to get the overall surface area. Next, multiply this by the number of layers in the laminate to find the total material surface area to be used. Lastly, multiply this by the weight listed by the material supplier or manufacturer to find the total reinforcement weight for the project. This final reinforcement weight (in ounces) can then be converted to pounds (by dividing by 16) and used to find approximate resin amounts, as described below.

For fiberglass laminates, resin quantities can be approximated by matching the same weight polyester or vinylester resin to reinforcement material when using woven fiberglass, or by adding nearly twice the weight resin to reinforcement when using mat fiberglass. Approximate resin weights are as follows:

Vinylester resins = 9lbs/gallon
Polyester resin = 8.8lbs/gallon
Epoxy (with hardener) = 8lbs/gallon

Notice that epoxy weighs slightly less than polyester and vinylester resins, and remember that carbon fiber and Kevlar fabrics have less density than fiberglass, so resin weight will take up more of the total weight in the composite (for the same volume fraction). As a result, plan to have at least 1.3 pounds of epoxy resin per pound of carbon fabric, or 1.6 pounds of epoxy resin per pound of Kevlar fabric. For example, to make a six layer laminate that is 2 square

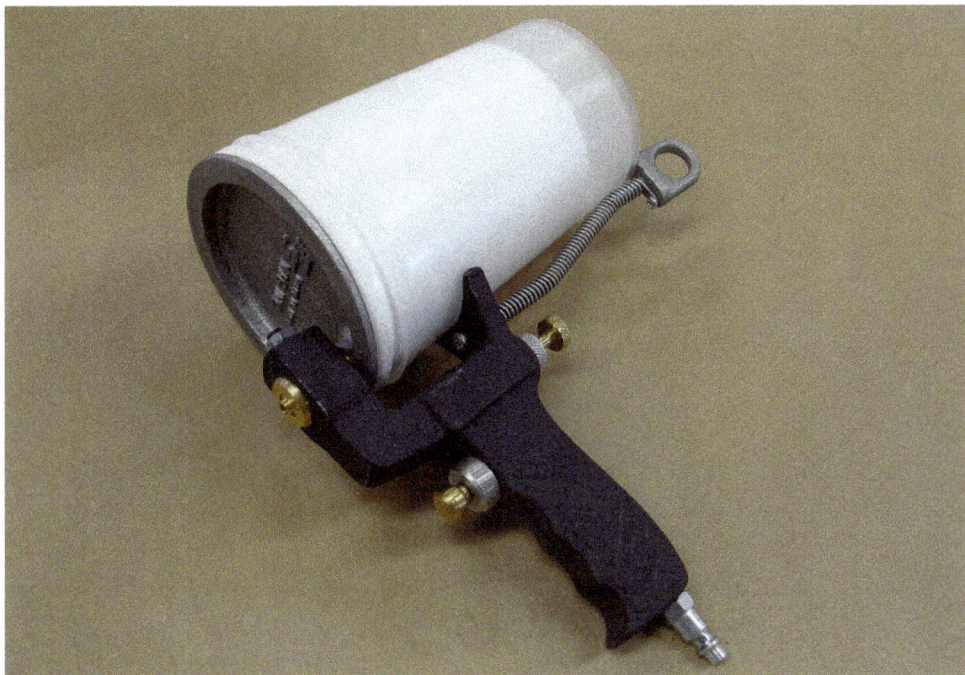

A gel coat "cup gun".

A cup gun is a fast way to cover large molds with tooling gel coat with excellent results (as shown here, being used to spray a small vehicle plug).

29

Cut all the necessary fiberglass mat reinforcements prior to starting the layup.

yards (or 18 square feet) using 1.5 oz fiberglass mat, the reinforcement materials would weigh a total of 162 ounces (or about 10.1lbs) and therefore require about 20.2lbs (or about two and a third gallons) of polyester resin. Based on their particular layup methods and experience, fabricators may use slightly more or less resin than this, but these calculated amounts should get most folks in the ballpark, so to speak.

Simple material and resin calculators are also available online - with an exceptionally handy one is located at www.fibreglast.com.

Plan enough time to do a correct layup, and have skilled help when creating large mold layups. As a rule of thumb, a properly catalyzed gel coat can take about an hour to cure after application before it is to the tacky state, and ready for layup. As far as planning for actual layup time, a single skilled worker can typically hand-laminate six layers of 1.5 oz fiberglass mat to a 10 square foot mold surface in about a six to seven hour period - or a moderately sized mold in a full eight-hour work day. Additional skilled workers assisting with a mold layup will typically bring the lamination time down by a couple hours per person - so a

After the tooling gel coat has cured to a tacky state, apply a thin coat of polyester resin or (as shown here) additional tooling gel coat over the entire form.

layup that requires one person six hours to complete could feasibly be done in only four hours with two people, and in roughly two hours with three people.

APPLYING TOOLING GEL/SURFACE COATS

Opaque tooling gel coats or surface coats make it possible to more easily see imperfections in a mold surface after mold layup and demolding. Polyester tooling gel coat and epoxy surface coat resin can be a little more expensive than laminating resins, making it tempting to use their unfilled, un-pigmented plain resin counterparts to form the surface of the composite tooling. However, it is a bit difficult to determine the location of actual surface flaws in a clear resin composite because bubbles in the composite are visible through the surface. By contrast, pigmented tooling gel coats and surface coats produce opaque surfaces that are easy to inspect because light reflecting on them will reveal any undesirable waves, high spots, or low spots. These specialized resins will also help hide any repairs made to the mold surface before or during its service life.

Apply a tooling gel coat (of polyester resin) or a surface coat (of epoxy) resin to the waxed plug and flange forms in

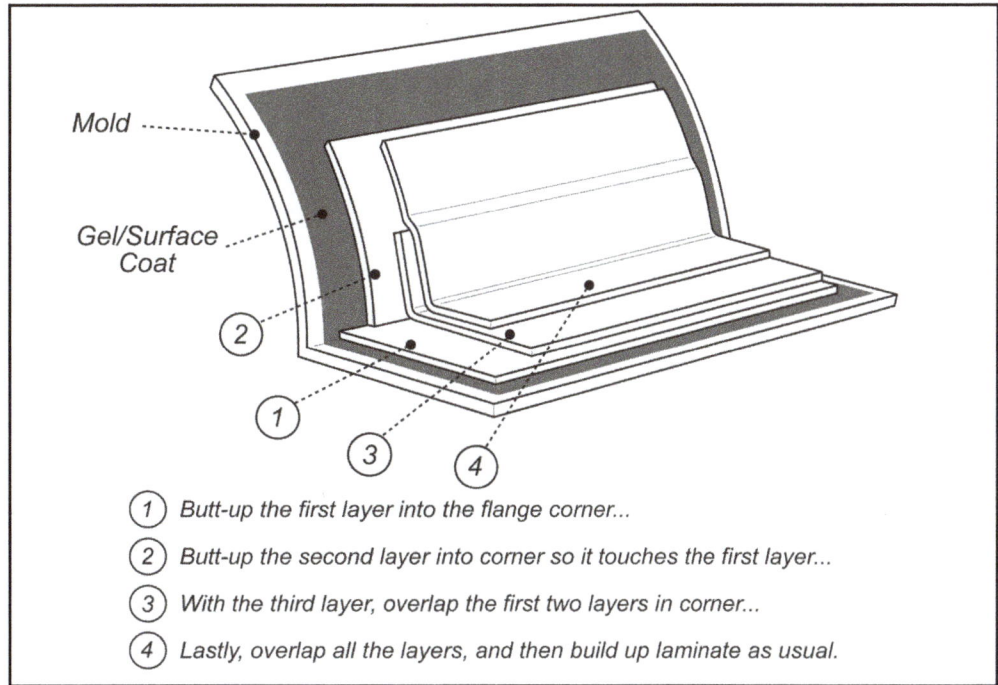

1) Butt-up the first layer into the flange corner...

2) Butt-up the second layer into corner so it touches the first layer...

3) With the third layer, overlap the first two layers in corner...

4) Lastly, overlap all the layers, and then build up laminate as usual.

The method for laying up tight corners at the flange.

Begin laying on the reinforcements. At the flanges, butt the fabric into the corner where the flange meets the plug...

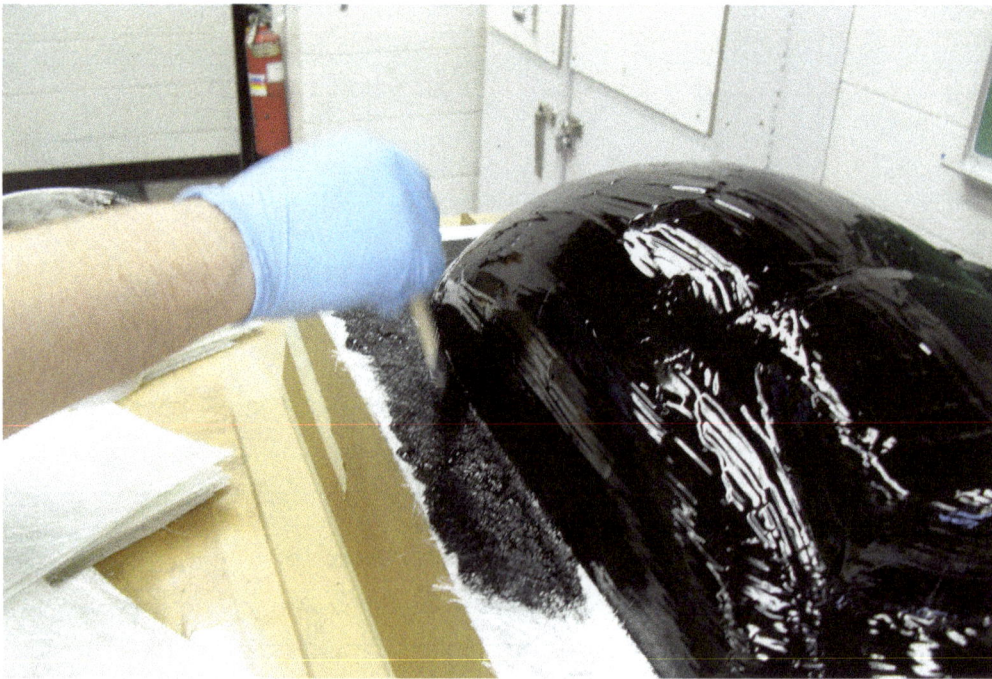

…saturate it with resin and wet out the adjacent side.

Apply another piece of mat, butting it into the corner as well.

thin, even coats. Thick coats can develop significant heat during the curing process, causing problems with release from the plug later on, especially where clay or wax fillets were created in seams and corners since they can soften or melt when heated by the exotherm of the curing resin. Multiple coats can be applied as needed to thicken the first gel/surface coat, but allow this first surface coat layer to cool and cure to a tacky state before applying additional coats. A typical gel coat thickness is between 12 and 20 mils (.012" and .020") and can be measured with a gel coat thickness gauge at the time it is applied. Keep in mind that a gel coat will shrink in thickness by about 20% to 30% during cure, so spray it slightly thicker to accommodate.

Most professional mold makers prefer sprayed application of gel coats, but for small molds or parts, a high quality brush will still work well and save some gun clean-up hassle. If using a spray gun, be sure to use a gun designed specifically for spraying gel coats, such as a "cup gun" that holds the gel coat in a removable paper cup and has an easily cleaned spraying orifice. A standard paint spray gun, though usable with gel coats, is not advisable because of the limited pot life of gel coats (which can cure in the gun and destroy it) and the extensive cleaning required immediately after spraying. If you must use a paint gun with gel coat, however, styrene monomer can thin it a bit so it will flow better through the gun - but be aware that adding more than 20% to 25% styrene (by volume) to the gel coat can make it overly brittle. Also, be sure to use goggles and a respirator when spraying gel coat since the atomized polyester resin can cause serious respiratory and eye irritation.

When using epoxy as a surface coat in place of tooling gel coat, keep in mind that thinning and spraying epoxy is not recommended. Epoxy surface coats are best applied with a brush, but can still yield excellent results. Whether creating your mold surface with either a gel of surface coat be sure to check for any pinholes in the final mold surface and fill them prior to putting the mold into service. For large holes, tooling gel coat or thickened resin may be needed for filling. Small pinholes, however, may simply require sufficient filling with modeling clay or paste wax - at least in the short run.

Dab resin into the fabric until it is completely wet-ted-out.

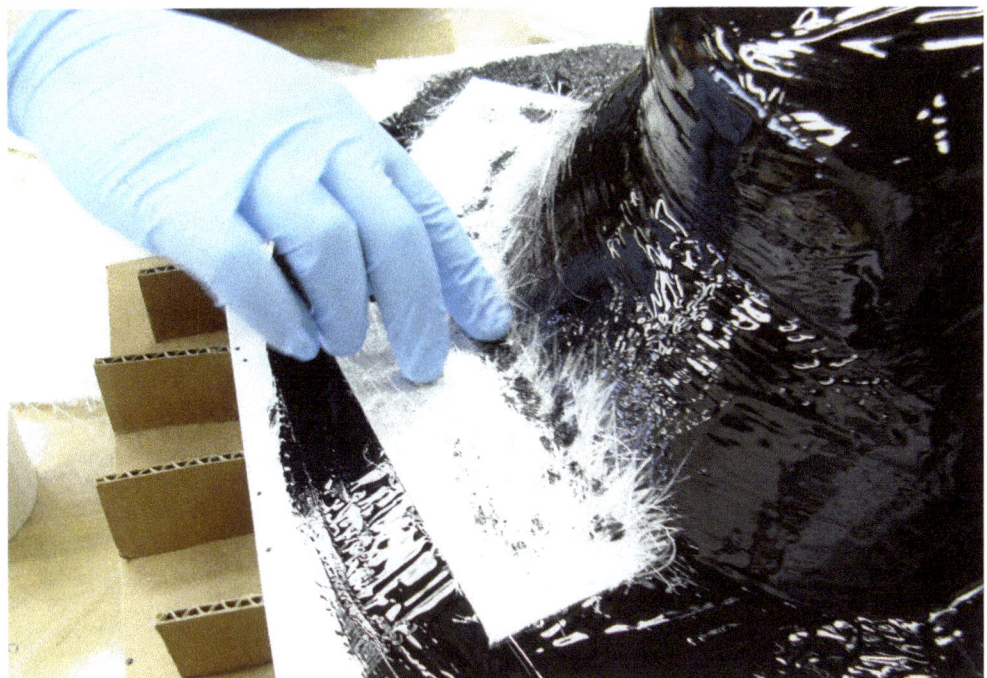

To create a strong bridge of material at the corner, tear rough edges in the mat, laying the rough edges so they lie over the corner...

33

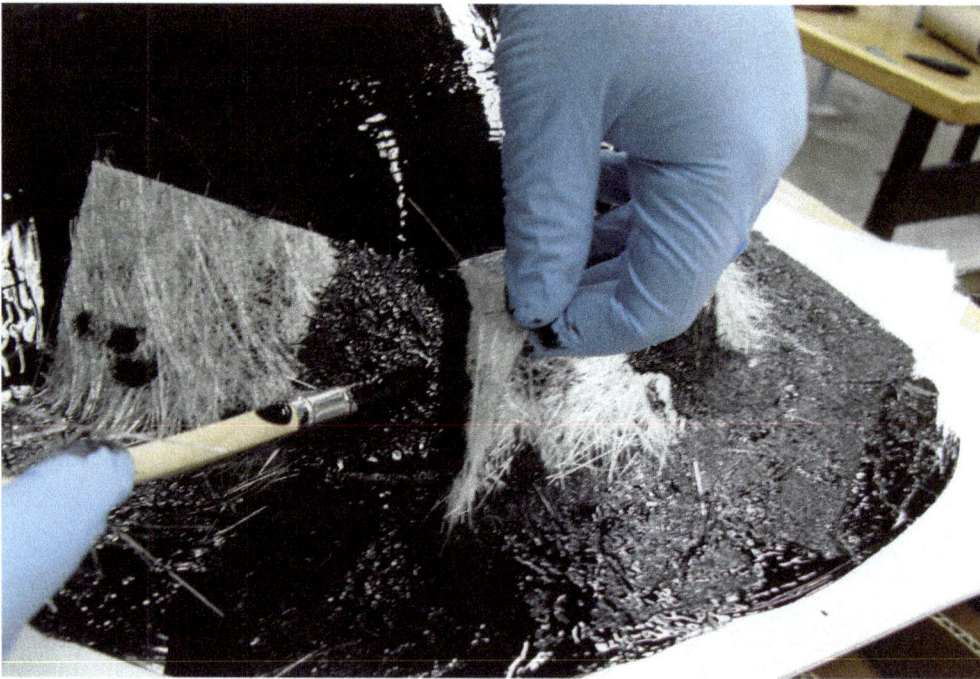

...and then do the same on the other side of the flange corner, wetting out the reinforcements completely with resin once they are positioned correctly.

MOLD LAYUP

Once the gel/surface coat has cured to a tacky state, apply a coat of resin over the gel coat, and then begin to add the reinforcement layers for the mold. For this particular tooling demonstration, the final parts molded in it will be between 1/16 and 3/32 of an inch thick. Therefore, this mold will be laid up to a thickness of between 3/16 and 1/4 of an inch thick.

Layup of these layers is the same as for a normal wet layup, where resin is brushed on the surface, fabric is laid over the resin, and additional resin is added where needed to completely wet-out the fabric. These steps should be repeated as necessary to build up the thickness of the laminate, again taking care to not build up the composite so thick that it gets warm enough to affect the plug or overheat the resin. If not performed at the right pace, a mold can be built up so quickly that the exothermic reaction of the resin will generate sufficient heat to actually start smoking - even with only a modest amount of catalyst or hardener! For example, 1.5 oz fiberglass mat laid up at room temperature with polyester resin that has been catalyzed to 1% MEKP (or about 5 cc of MEKP cat-

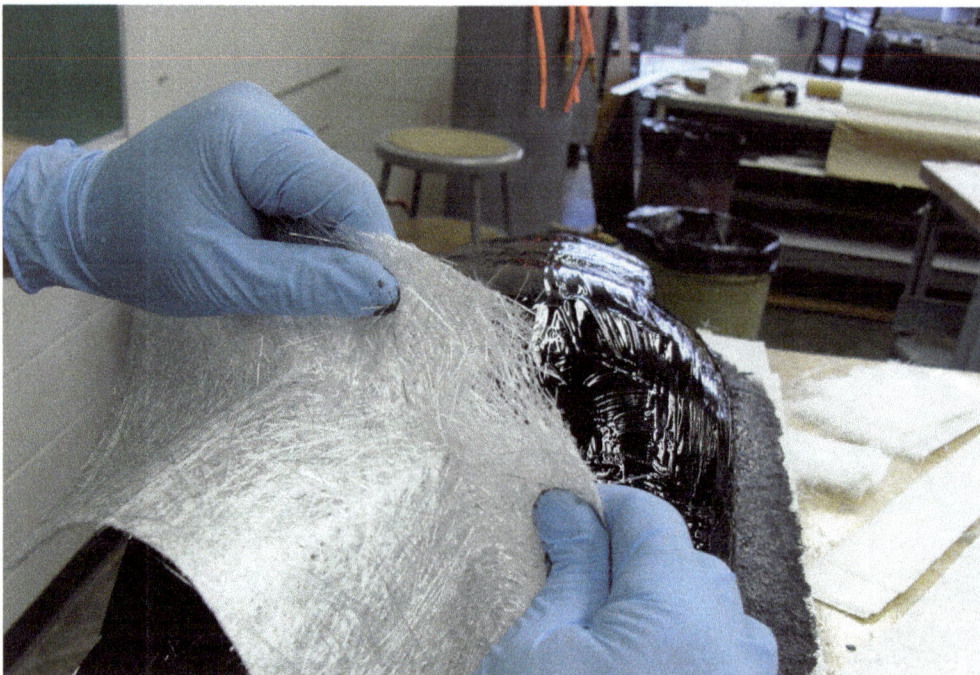

Tear the edges of all remaining mat reinforcements so that edge voids will not form in the laminate.

alyst per 16 oz of resin) will produce noticeable heat when it is laid up at a rate of only one layer every ten minutes. A layup performed any faster or catalyzed more than this will very likely overheat the resin. For situations requiring fast laminate buildup, less catalyst may be helpful, as long as there is enough catalyst present in the resin to sufficiently cure it.

In areas where there are tight corners (such as where the plug meets a flange), create a fillet of thickened resin (filled with an appropriate thickening agent such as colloidal silica) in the corner with a gloved finger, or butt-up strips of reinforcement on either side of the corner, then bridge it with additional layers of fiberglass mat to reinforce it. Continue with this build up of the mold until the flanges are sufficiently thick and reinforced.

Most resins will exhibit some amount of "creep" (or permanent deformation of the material when placed under prolonged load), so composite tooling should be reinforced in a manner that minimizes this tendency as much as possible. To avoid mold warping from creep, some fabricators will reinforce the backside of their molds by laminating on ribs of wood, polyurethane foam and composite, or

For the areas of the mold registration, cut the fabric so it will build up over the detail in successive layers.

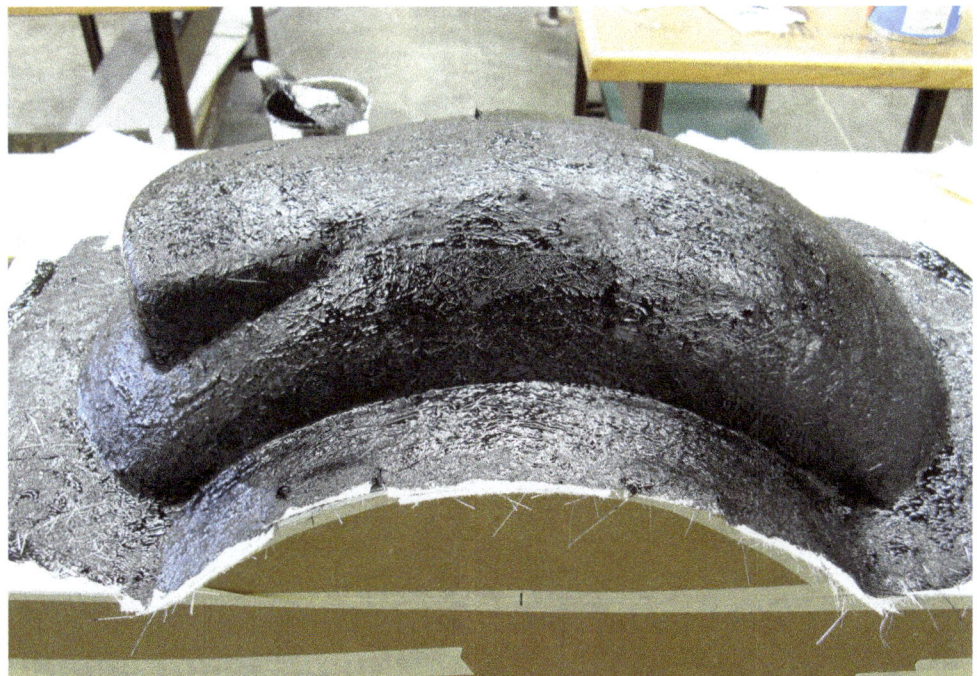

Build up the layers of the mold in succession until the lamination schedule is complete.

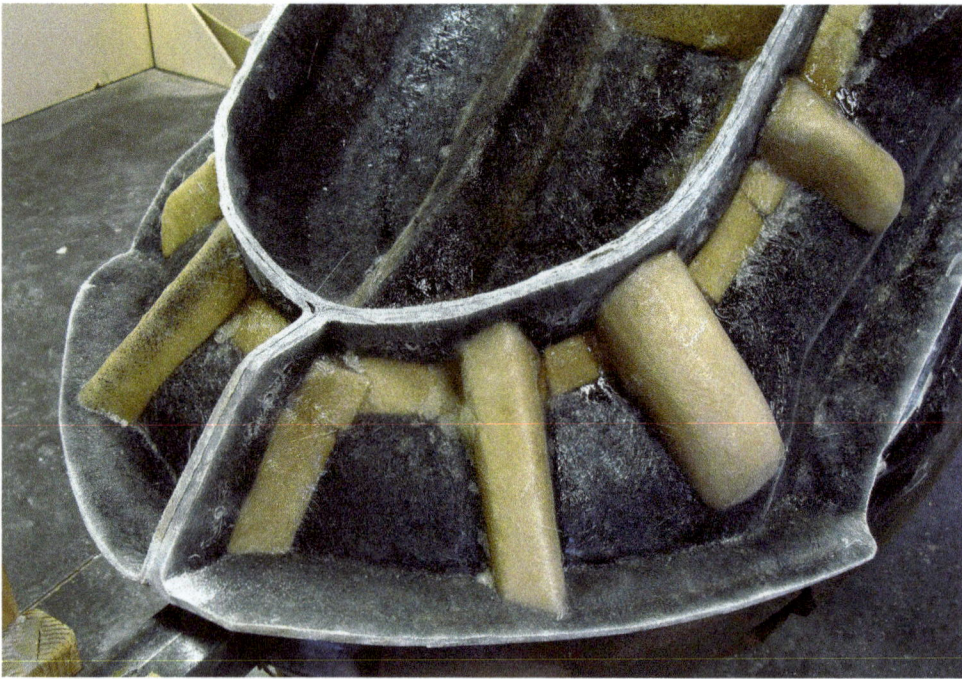

Reinforcement ribs can be as simple as polyurethane foam adhered to the mold with thickened resin and then laminated over with fiberglass (as shown here on a multi-sectioned mold where support was critical).

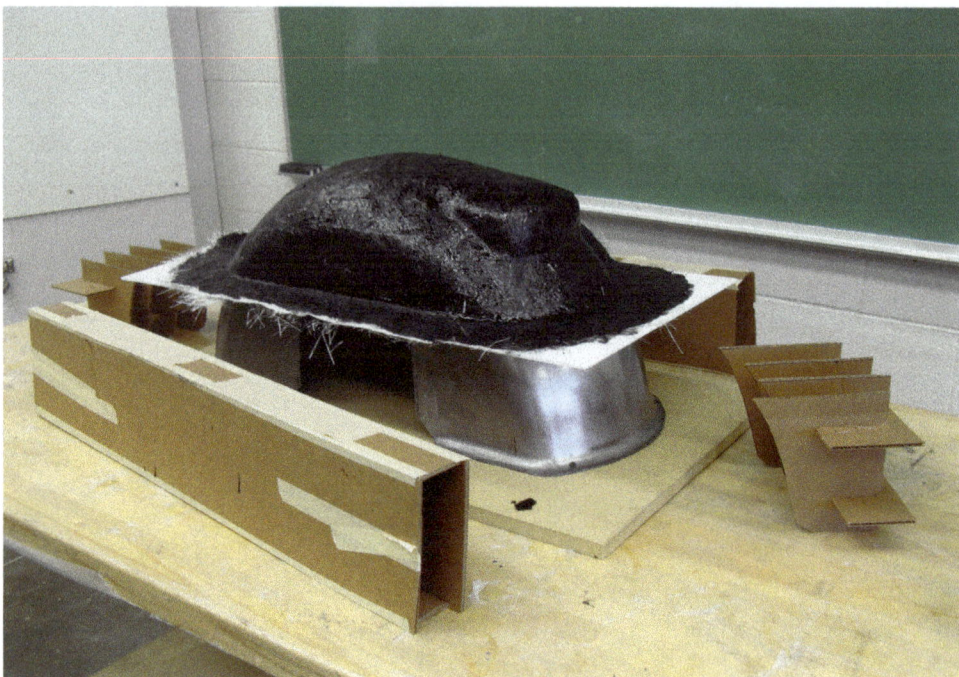

Once cured, remove the flange form supports and flange forms.

even angle iron or steel tubing. Carbon fiber/epoxy reinforcements added to the laminate can also slow the effects of creep in a mold because the inherent properties of carbon/epoxy laminates help them hold their formed shape very well over time.

After lamination, allow the mold to cure completely before removing it from the plug or forms. With most common epoxies and properly catalyzed polyester and vinylester resins, wait at least 24 hours before demolding the composite tooling. Fully cured molds will demold more easily and won't be at risk for permanent deformation.

Demolding and finishing procedures for composite tooling are the same as for other molded composites. When demolding composite tooling, take care to minimize damage to the plug, just in case it will be needed again for mold repairs in the future. Mold flanges can be trimmed most easily using a cutoff wheel (with either an electric or air-powered rotary tool) and rough, stray fibers can be removed from the back of a mold using a file, sandpaper, die grinder, or disc sander. See *Composites Fabrication Handbook - 1* for demonstrations on demolding and finishing of composite tooling.

MULTI-PART MOLDS

For molds that must accommodate complex geometries, multiple removable mold sections may need to be designed into the tooling. Mold flanges can be created between such multi-section molds, but need to incorporate "mold registration" marks and alignment holes to ensure correct fit between each mold section.

Construction of a multi-sectioned mold requires building the flange forms and mold sections in multiple steps. Flanges on multi-section molds are typically created by first building an initial flange at the parting plane using the same methods for flange formation explained above. If perfect alignment is needed between mold sections, it is advisable to construct registry marks out of clay (or other appropriate material) and attach them to the flange forms so they will transfer into the newly formed flanges during layup (as demonstrated in this chapter). When the flange forms are removed, they will reveal the flange registration marks onto which additional mold flanges and sections can be laid up and keyed into.

Make sure to apply wax and PVA to the flanges before beginning layup of any subsequent mold sections. Beware of stray fibers

Position the mold upside down and prepare it for layup of the second (removable) mold section.

Carefully remove any clay remaining in the mold registrations. Use naphtha or mineral spirits and a rag to remove all clay residues.

Wax the plug and flange.

or frayed edges on the flanges; if these are not sealed (using packing tape or polyethylene flash tape), resin from the new layup may bond with them and fuse the two mold sections together. The second matching flange and mold section are then constructed to meet the previous flange, ensuring perfect alignment of the mold sections.

After this new mold section has cured, drill mold alignment holes straight through the mating flanges before separating the molds and removing them from the plug. These holes can be used during future layups to ensure good alignment of the mold sections, with bolts and washers tightened through them to hold the sections securely together. Trim and smooth all resulting edges for safe handling of the mold during subsequent use. Keep in mind that additional mold sections can make sealing the mold a bit more difficult for vacuum bagging applications because the outer edges of the parting plane must be closed off with sealant tape to avoid air leakage.

CARE OF COMPOSITE TOOLING

Whether a piece of composite tooling is to be used for extensive periods of time or just stored between infrequent layups, it must be treated with

Add a new flange form to the top of the mold section area, and apply a bead of clay to any remaining seams.

care to ensure a long service life. Occasional cleaning with water and mild soap will help keep dust and other contaminants off the mold surfaces and out of any composites laid in the mold. If uncured resin comes in contact with the mold, solvents may need to be used to remove them. Always use the least harsh chemicals and cleaning methods possible before attempting to use more severe ones that may damage the mold. In the case of spilled resin, wipe up as much of the resin as possible with a rag, and then use a citrus-based soap to clean up remaining residue. If more cleaning power is needed, use sparing amounts of acetone or lacquer thinner to finish the job.

Over time, a thin film of cured resin may build up on the surface of a mold when gel or surface coats are not used. This may be apparent by a faint ghost-like fiber pattern that is visible on the mold surface. To remove this film, it will be necessary to apply cutting/rubbing compound onto the mold's surfaces, and then to buff and polish it until it has regained its original gloss.

If composite tooling needs to be stored for long periods of time,

Cover any stray fibers with clear packing tape to prevent resin from seeping into them and locking the two mold sections together.

Apply the tooling gel coat to the surface of the removable mold section and then allow it to cure to a tacky state.

Add additional resin and reinforcements to the mold section, building it up as necessary to complete the lamination schedule.

preparation should be made to ensure that it will be stored in such a way as to maintain its proper shape until used again. Mold reinforcement can help significantly in this regard, but some fabricators will also store the mold over the original plug, or fabricate a special polyurethane plug (using two-part mixed urethane foam poured into the waxed mold) to provide support to the mold during storage. Never store heavy objects or shop materials on or in composite tooling, as these will inevitably deform the mold or mar the mold's surfaces. Store composite tooling by supporting its flanges rather than by letting it rest on any reinforcement ribs laminated to its backside; pressure on these ribs may cause localized waves in the mold surfaces over time.

Lastly, it is wise to apply a good coat of wax on a mold before storing it for any long period of time. This will make later cleanup and reuse of the mold much easier, especially if dust, debris, or other messy materials find their way onto the mold. In short, treat composite tooling with a healthy respect for the time and effort that was put into its creation.

Following these composite tooling fabrication

Once cured, drill holes through the flanges at the removable mold section (for fastening with nut and bolt hardware) prior to demolding to ensure perfect mold alignment.

and care guidelines will help produce molds that are useful for forming hundreds of parts. Proper use and storage of these molds will then save the considerable cost of replicating a mold while significantly maximizing its lifespan and providing a high-quality, useful mold.

CHAPTER CONCLUSION

The steps for creating composite tooling are very similar to those found in other wet layup procedures. Building a suitable plug and attaching the flange forms prior to mold layup can be the largest challenge in the fabrication process, but can become second-nature with practice. A well-designed composite tool will provide years of service in molding high quality composite parts - paying for itself many times over when compared with other types of mold-making processes and materials.

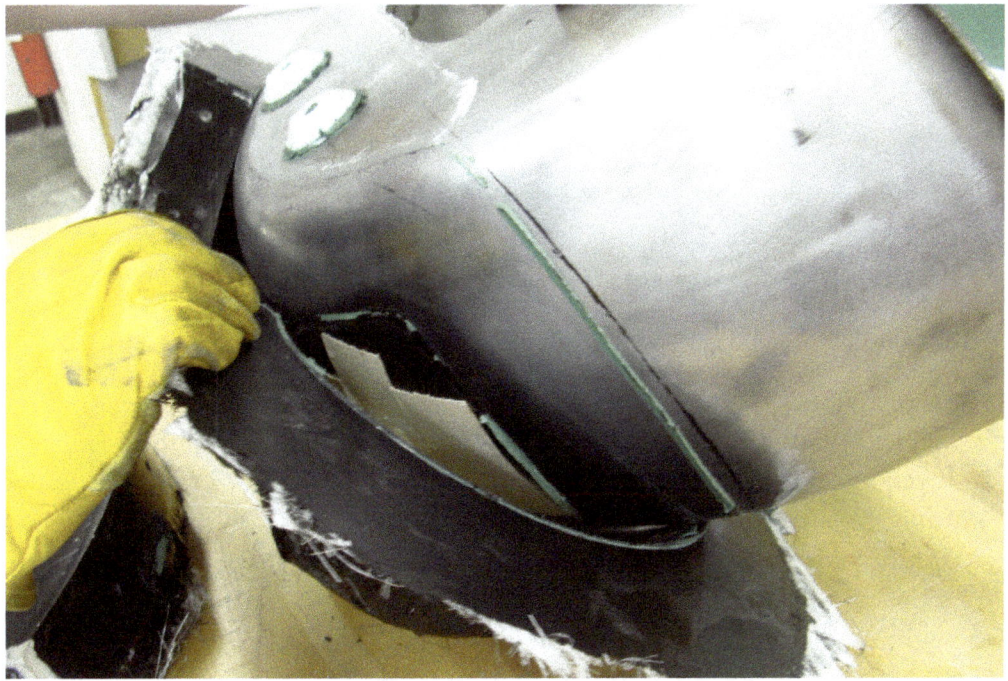

Using gloved hands and plastic or wood wedges, demold the composite tooling…

…and trim and finish it completely before putting it into service (as shown in Handbook #1).

Chapter Three

Compression Molding Techniques

A High-pressure Laminate Made Simple

CHAPTER INTRODUCTION

When it comes to optimizing the strength and weight of a smooth-surfaced composite, "compression molding" is probably the most straightforward means of fabrication. Compression molding is performed by placing the uncured laminate in a closed mold (or "matched mold") and applying pressure to con- solidate the laminate and press out excess resin. This method is used in industry for mass-produced parts where correct resin/fiber ratios and surface quality are critical. This chapter will show how to use a simple version of this process to produce smooth, high quality panels that can be trimmed and used for a variety of purposes.

Compression - molded panels can produce a variety of high quality flat composite products - from lightweight, rigid panels, to automotive trim, to simple guitar picks.

COMPRESSION MOLDING BASICS

Industrial scale compression molds are typically made of metal and compressed using large industrial presses. Such molds are frequently equipped with the ability to cure laminates quickly by using heating elements, heated fluid lines, or even inductive heating. Because of the high pressures and fast part cure possible with compression molding, it is ideal for quick turn around of high quality production parts, and is the most widely used means of manufacturing high-volume "Class -A" surfaced composite body panels in the automotive industry.

Due to the excellent surfaces possible with compression molding, aesthetic surfacing steps, such as applying polyester gel coats or epoxy surface coats, become less important. In fact, the pressures applied through compression molding make these specialty coatings practically useless during processing. With high compressive molding forces, reinforcement fibers will either be pressed through the coating (when in its tacky state), or the coating itself can fracture (if it has previously cured). Regardless, the final surfaces of the compression molded composite alone will need very minimal filling or surface prep and are optimal for painting or clear coating.

Two of the biggest limitations that small shop or home-bound fabricators run into with compression

molding are that it works best for relatively simple shapes, and that it usually requires expensive, specialized equipment and molds. Shapes that can be molded with this method include those with shallow depth, minimal surface curvature, no undercuts, and no vertical features (such as ribs or fins). Parts that require more complex shapes can be molded using other means—like some of the additional fabrication techniques outlined in this book.

Compression molding's limitations to small fabrication shops are further magnified by the cost and complexity of setting up a typical system. Metal molds and presses (whether mechanical, hydraulic, or pneumatic) are required for nearly all compression molding applications because of the high pressures necessary to pro-

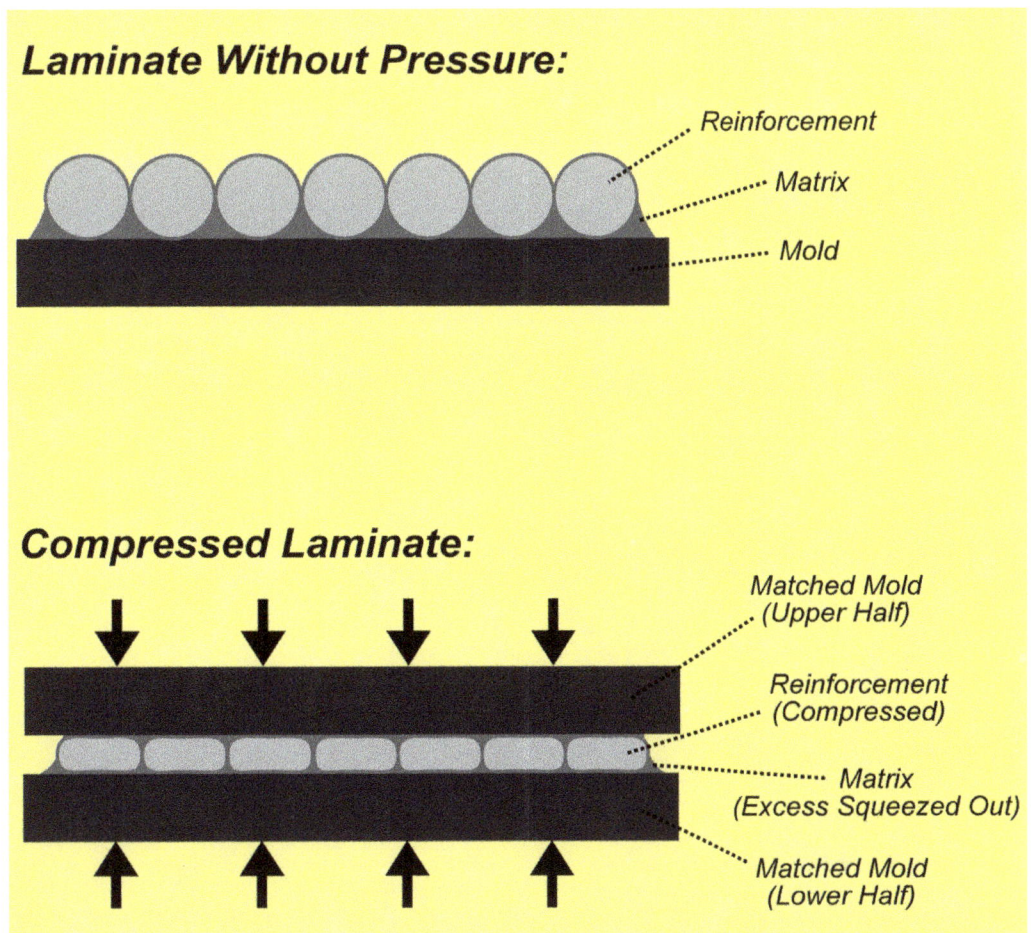

The effects of compression on a composite's resin content and thickness.

duce quality results. Yet, despite these apparent limitations, simple compression molded panels can be easily produced using common tools and rigid sheet materials available at the local hardware store. These panels can be fabricated quickly and are useful for any application that requires a laminate that is smooth on both sides.

For the demonstration in this chapter, a PVC (polyvinylchloride) foam material will be added as a "core" for greater rigidity in the final laminate (though it could be left out of the layup without problem). "Cores" are materials added

between laminates to create stiff "sandwich" structures—so named because they are stacked similarly to a sandwich. Although it is out of the breadth of this book to describe the proper design of sandwich composites, suffice it to say that there is an exponential correlation between the thickness of a cored laminate and the strength and stiffness it exhibits. Sandwich laminates are similar in concept to the I-beams used in large scale constructions—the top and bottom sections bear the majority of the load on the structure. The middle section of the structure simply holds the top and bottom at a fixed distance from each other so they can work effective as a unified load bearing structure. The greater the distance between the top and bottom sections, the stiffer the sandwich laminate. For example, a four ply laminate, if separated by a core to twice it's original thickness (with two plies on top and two on the bottom), would be 7 times as stiff and 3.5 times as strong. If that same top and bottom laminate (still at two plies each) employed an even thicker core to take it to 4 times its original thickness, it would have 37 times the stiffness and over 9 times the strength—with very little increase in weight! Needless to say, cores help make good composite structures.

A variety of materials can be used as cores, but the most effective ones are those that are lightweight, have good compressive strength, and are readily bonded with the resin used in the laminate. Some common cores include foam (urethane, PVC, expanded/extruded styrene), wood (end-grain balsa, plywood), and honeycomb (Nomex, aluminum, and other). For applications requiring low cost and low performance, wood cores work well, while high performance laminates typically use honeycomb cores. Moderate price and performance laminates tend to benefit well from foam cores. The core processing techniques shown in these next two chapters can be used with almost all types of cores that are employed in a wet layup.

Compressed Panel Lamination Schedule

Resin System: Epoxy
2 Layers - 5.5 oz. Carbon (4HS Satin)
PVC Foam Core
2 Layers - 5.5 oz. Carbon (4HS Satin)

Materials and Supplies

Melamine-faced Particle Board (3/4" Thick)
Mylar Film
Polyethylene Film (3 Mil)
Clear Packing Tape
Colloidal Silica Thickener
Mold Release/Parting Wax
Mixing Cups and Sticks
Spreader and Disposable Gloves
Clean Rags/Cloths
Scissors and Ruler
Wood Clamps

The compression molding lamination schedule, along with materials and supplies, used in this demonstration.

STEPS TO SMALL-SCALE COMPRESSION MOLDING

To get started, cut all reinforcements to size and stack the fabric as desired for your application. If including a PVC foam core in the lamination schedule, place it between the plies of fabric as it will be laminated in the layup. PVC foam cuts very easily with a sharp utility knife and, if thin enough, can be shaped slightly with a heat gun or oven to conform to simple mold curvatures (as shown in the next chapter).

To create a robust, small scale compression molding system, thick boards are recommended for the mold since they will flex less under pressure, providing more even pressure distribution over the laminate. Start by preparing melamine-faced boards as you would any mold surface, cleaning the surfaces as needed, sealing off the porous edges with packing tape, and then applying three to four coats of wax, followed by PVA (polyvinyl alcohol), as shown in Handbook – 1. Mylar film can also be added over the mold faces for a glass-like smooth surface.

Apply resin to the mold faces with a spreader, and begin the layup procedure, working resin into each layer of fabric as needed for full wet out. Once the core

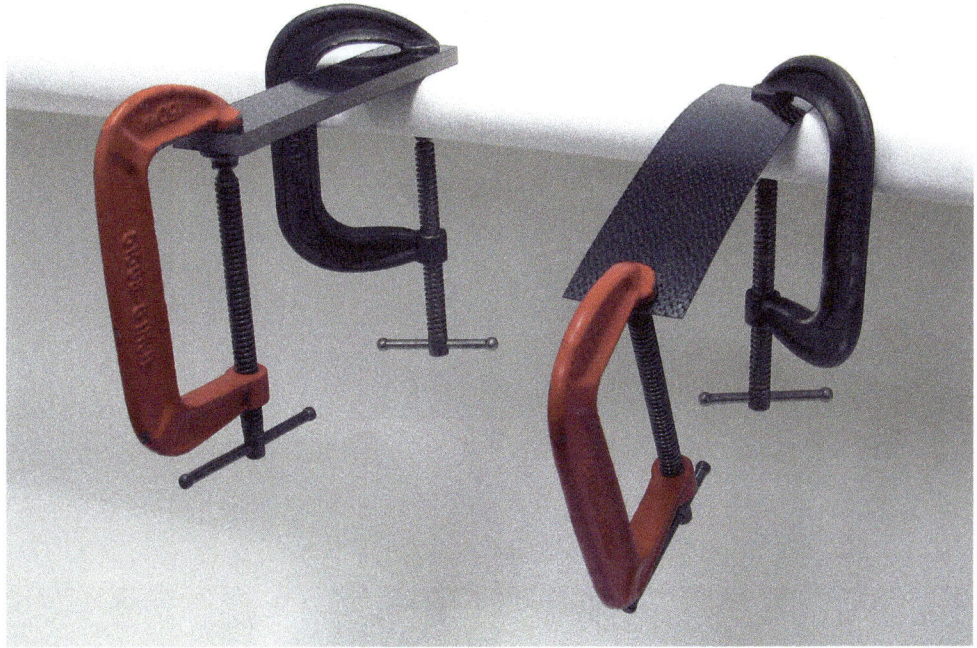

The stiffness of a sandwich laminate is greatly improved by the addition of a core, as shown by these two test samples of the exact same weight.

Start the layup by cutting all fabric materials.

Continued on page 50

45

Cut the PVC foam core using a utility knife and a ruler.

Mylar film can simply be placed over the board...

Mylar film (if desired for a smooth laminate surface) can be cut with scissors or a utility knife and ruler.

...and then taped at the edges. No additional release agents are needed over the Mylar film.

Apply packing tape to any porous edges on the Melamine-faced boards.

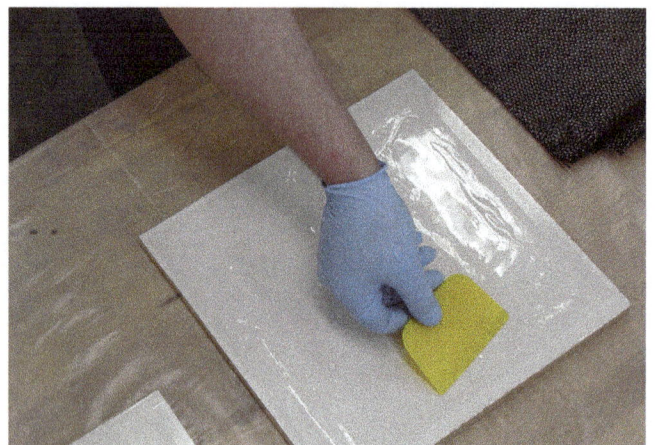
Spread the resin evenly over the surface of the bottom mold half to wet-out the laminate from below...

...and then add the first layer of reinforcement fabric.

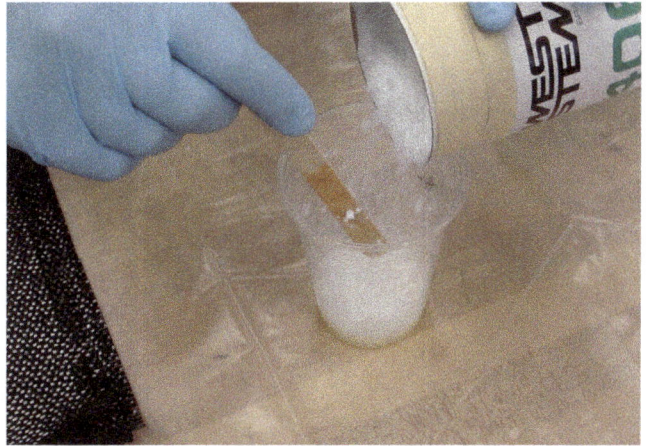

If adding a core, mix up some thickened resin by stirring colloidal silica thickener into the resin until it has a peanut butter consistency. Remove all lumps from the resulting resin paste.

Spread the resin through the fabric, and add more as needed to saturate the fibers.

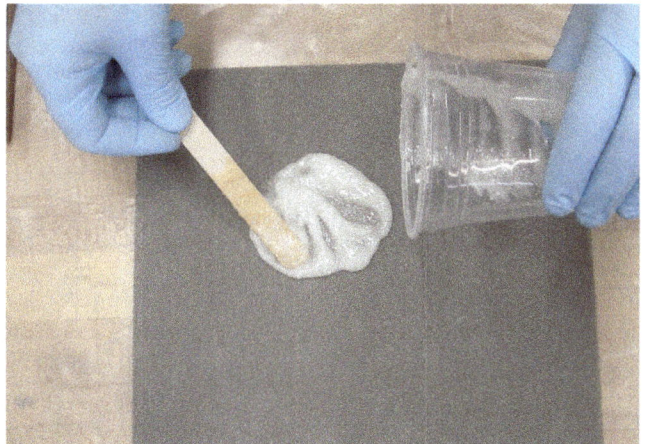

Put some of the resin paste over the face of the core...

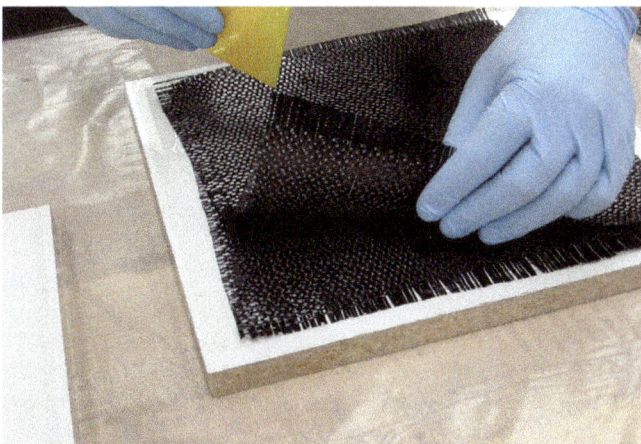

Additional plies can be added and wet-out with resin, as needed, for more reinforcement.

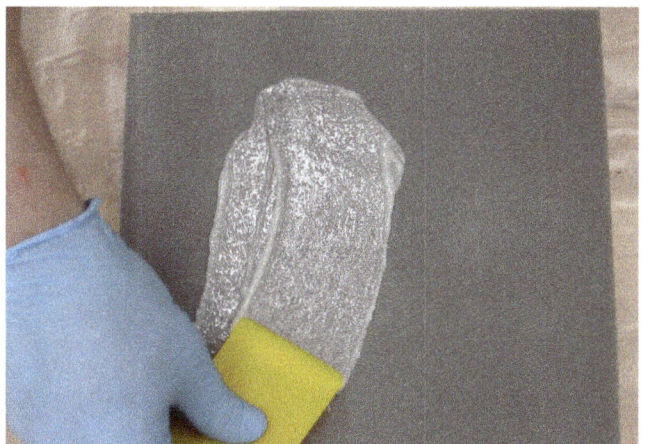

...and use a spreader to distribute the paste over the porous surface...

...until the surface of the core is completely covered.

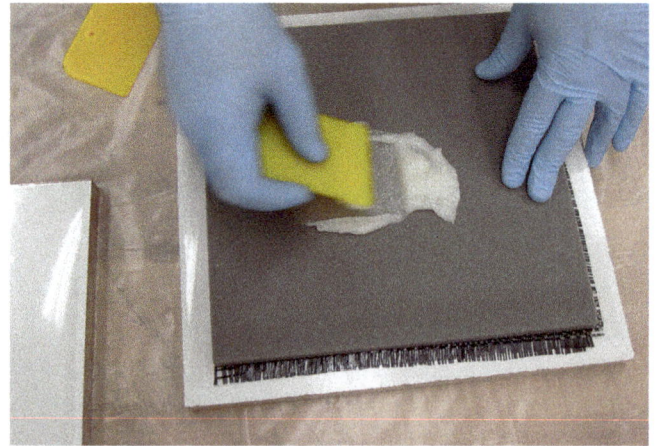

Add more resin paste to the open face of the core to prepare for the top laminate layers.

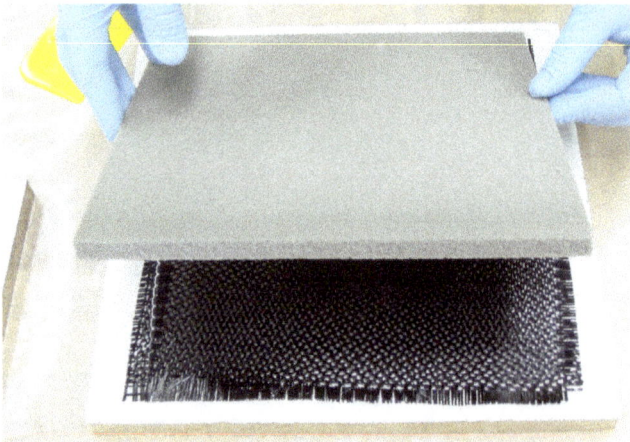

Place the core resin-side down on the laminate...

Pour more liquid resin onto the core to wet-out the subsequent laminate from beneath...

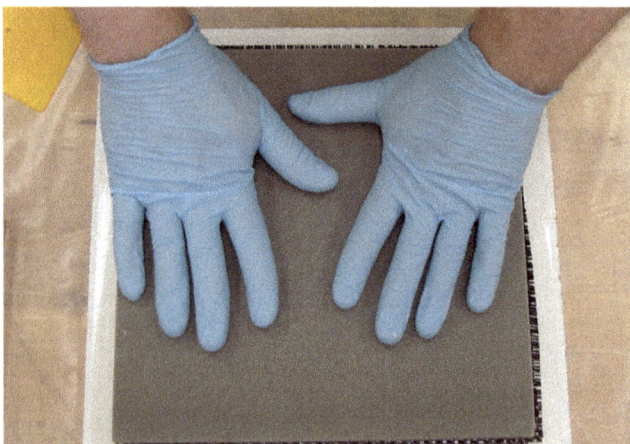

...and apply firm pressure to ensure a void-free interface between the core and the laminate.

...and level it out with a spreader.

Apply the remaining layers of reinforcement fabric over the core…

The mold set should have a equal distance between the boards on all edges.

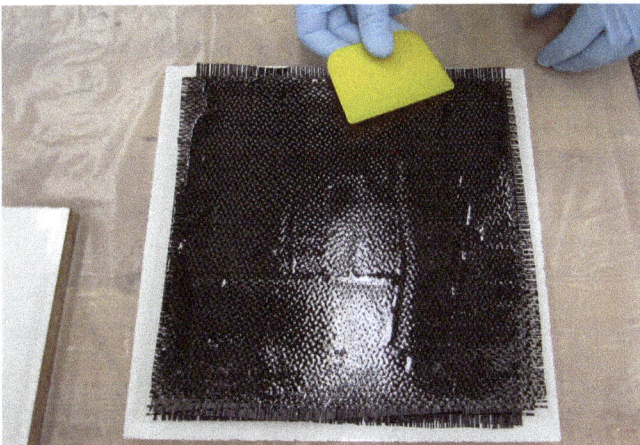

…and fully saturate them before closing the mold.

Place the molds and laminate in a polyethylene bag (made from film with taped edges).

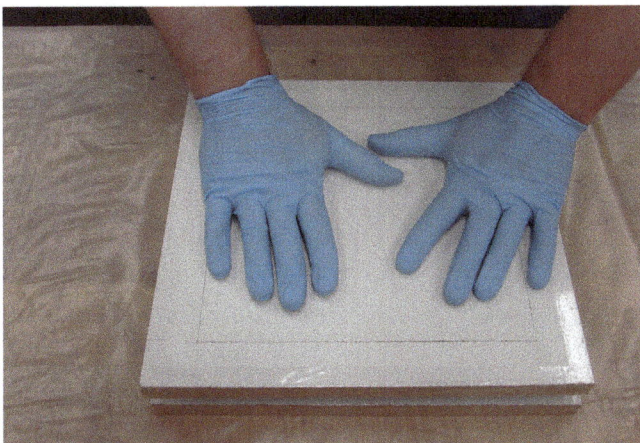

Place the top half of the mold on the laminate and apply firm pressure. Avoid any lateral sliding of the boards or laminate.

Apply pressure to the molds by using wood clamps. Clamps like these should be screwed down evenly for uniform pressure.

Add multiple clamps around the molds…

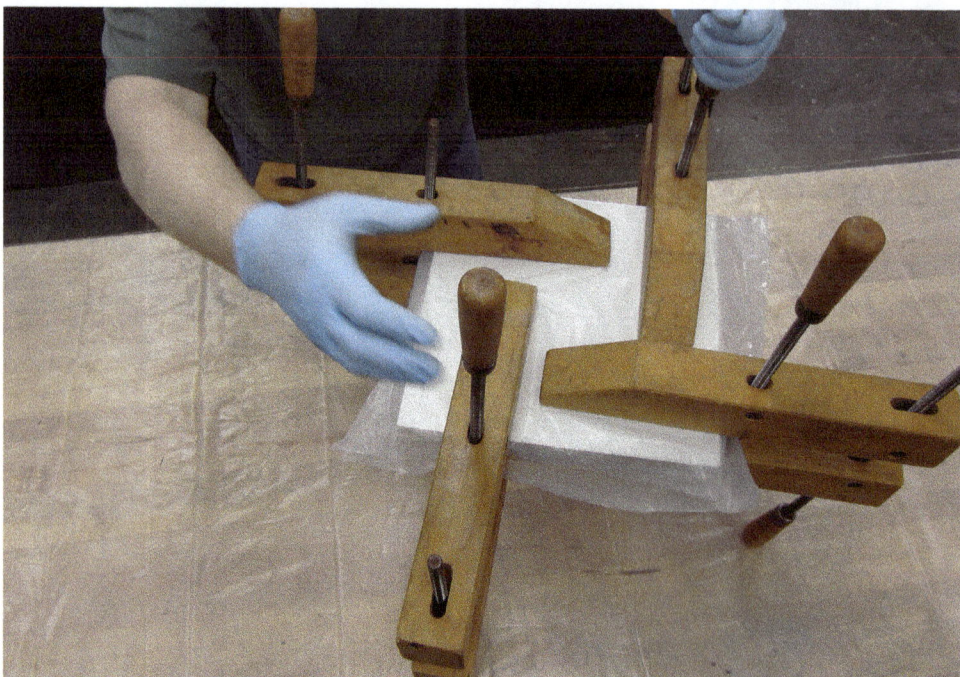

…until they are distributed well on the molds.

layer is ready to be laid on the mold, use a thickened resin (with colloidal silica filler added to make a paste) to close the pores of the foam and place it on the laminate. This will help increase the bondable surface area between the core and the composite and minimize delamination between the two. Continue with the final layers of composite over the core, as required by the lamination schedule.

Once the lamination is complete, close the mold by placing the second mold board over the laminate. Place the entire mold set in a plastic bag or between polyethylene film to help contain any resin spills. Next, compress the mold boards together by attaching wood clamps to the wide faces of the mold, evenly spacing them near the center and sides of the mold to provide balanced pressure on the laminate. Through this process, excess resin will squeeze out of the laminate, so take precautions to keep resin from spilling onto the threads of the clamps (which would destroy their usefulness for any future projects) or onto any surfaces that may otherwise be damaged by the resin.

When the resin begins to gel, apply additional clamping pressure by twisting the clamp screws even more. This will help close any spaces left at the mold faces after the liquid has been squeezed out, and will ensure a smoother final surface.

After the laminate has fully cured, it can be easily demolded. Remove the clamps, and pull or pry the molds apart (being careful not to scratch the mold surfaces) using a wood or hard plastic wedge. To demold the laminate itself, insert a wedge under it and release the panel from the mold face. Avoid pressing the wedge between the laminate plies or core, as they could separate and weaken the panel. The final panel should be very rigid and have an excellent surface finish. It can now be trimmed as usual and put into service.

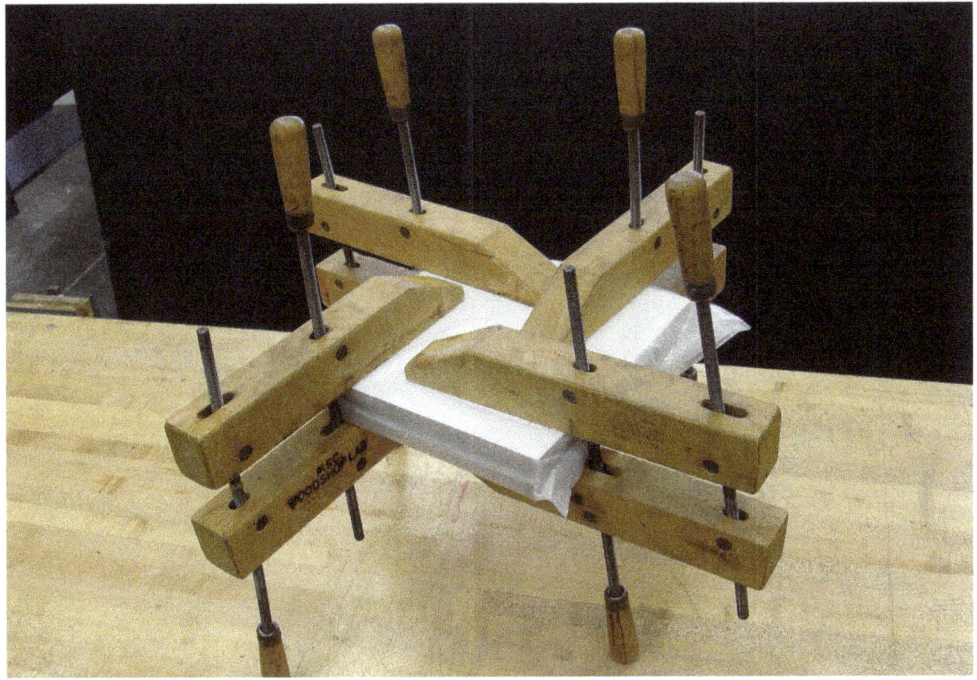

Allow the laminate to cure, tightening the clamps even more as the resin begins to gel.

CHAPTER CONCLUSION

High quality panel laminates and shallow forms can be made very easily using average wood-working tools and common materials. Though these materials may not have the same scale and overall capabilities of industrial-grade molds and presses, the results can still be quite reasonable and useful for the home-brew or small-fabrication shop applications.

Once cured, remove the clamps…

...remove the mold from the bag...

...and insert a wedge between the boards to remove the laminate.

Remove the top mold...

...and use the wedge to remove the laminate from the bottom mold.

Trim the laminate panel as needed. Panel trimming methods are explained in-depth in Handbook -1.

The resulting laminate should have excellent fiber consolidation and minimal excess resin within it.

Compression molded panels can be used for a variety of projects that require flat composite materials - like these fiberglass and carbon fiber knife handles (courtesy of Brad Southard, www.southardknives.com).

Chapter Four

Vacuum-bag Molding Techniques

An Aerospace Process, Done in Your Garage

CHAPTER INTRODUCTION

"Vacuum-bagging" is an advanced composite molding technique that uses a pump to draw a vacuum over the composite laminate, allowing even atmospheric pressure to compact the laminate against the mold face and improve its performance. This method of applying pressure against the laminate is not limited to flat parts alone (as with compression molding), but can be used just as well with complex shapes. This chapter explains the equipment and methods used to perform this helpful, composite-optimizing process.

VACUUM-BAG MOLDING BASICS

Applying a vacuum to an uncured composite provides several benefits to the final laminate.

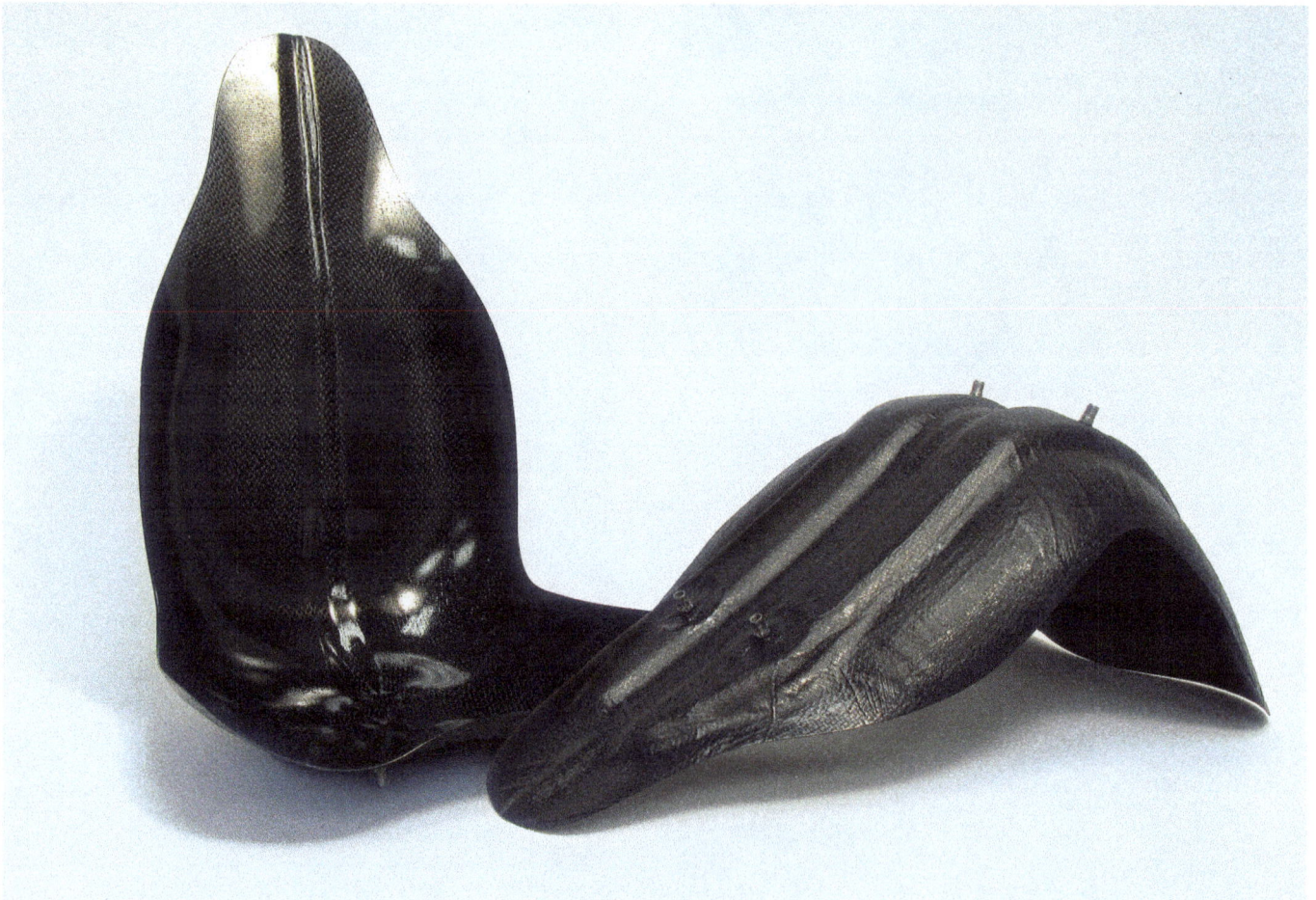

"Vacuum bagging" is a common means optimizing a composite laminate—as demonstrated in this chapter by the fabrication of a carbon fiber go-cart seat.

First, a vacuum can remove air and voids from the laminate, both from within it, and where it meets the mold faces. Voids in a resin matrix create weak points from which fractures can form and cause delamination and failure of the composite - so minimizing these voids helps increase the laminate's reliability. Evacuating an uncured laminate also causes atmospheric pressure to compress it and force out excess resin into absorbent processing materials to eliminate the unnecessary brittleness and weight that the surplus resin carries with it. This same pressure squeezes and consolidates the laminate plies together and compresses the reinforcement's weave against the mold face. Consequently, the higher the external air pressure, the more the laminate is pressed against the mold, and the fewer surface imperfections there will be where the laminate meets the mold.

To illustrate the amount of force that can be applied using this method, a simple 12" by 12" panel (144 square inches of surface area) at sea level (about 14.7 psi of atmospheric pressure) under a perfect vacuum would be evenly compressed with about 2117 pounds of force - the equivalent force of a compact car's weight concen-

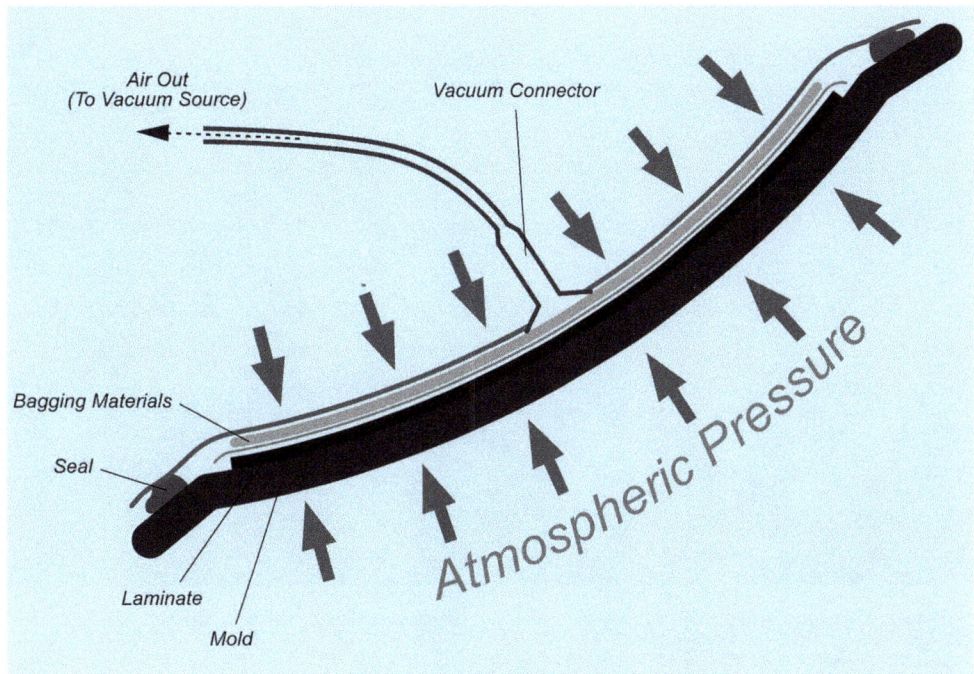

Vacuum bagging allows ambient air pressure to compress the laminate while drawing out excess resin.

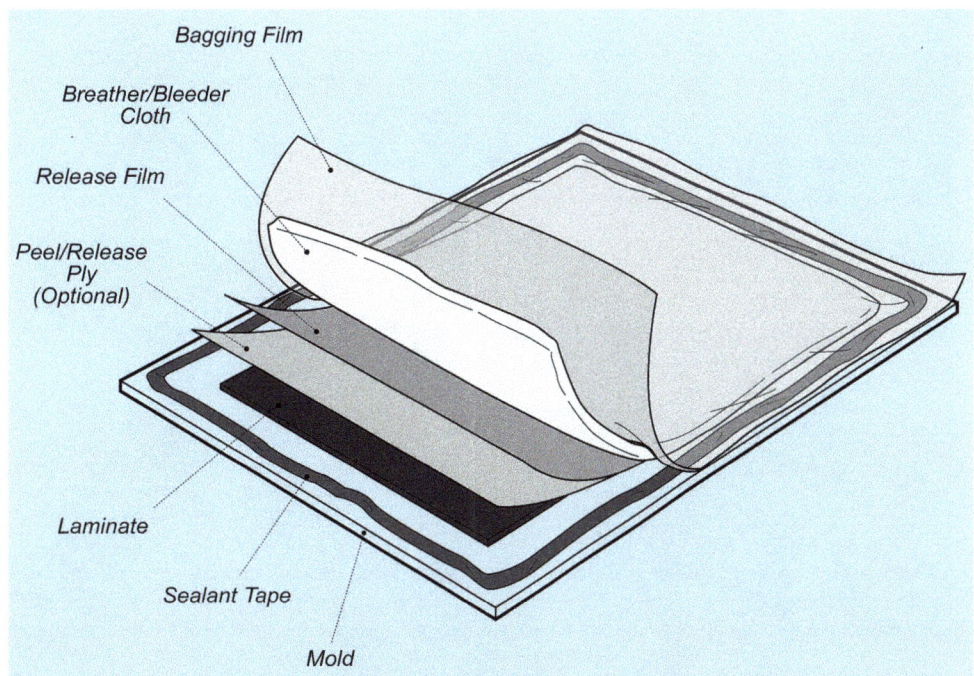

The layup order of vacuum bagging materials.

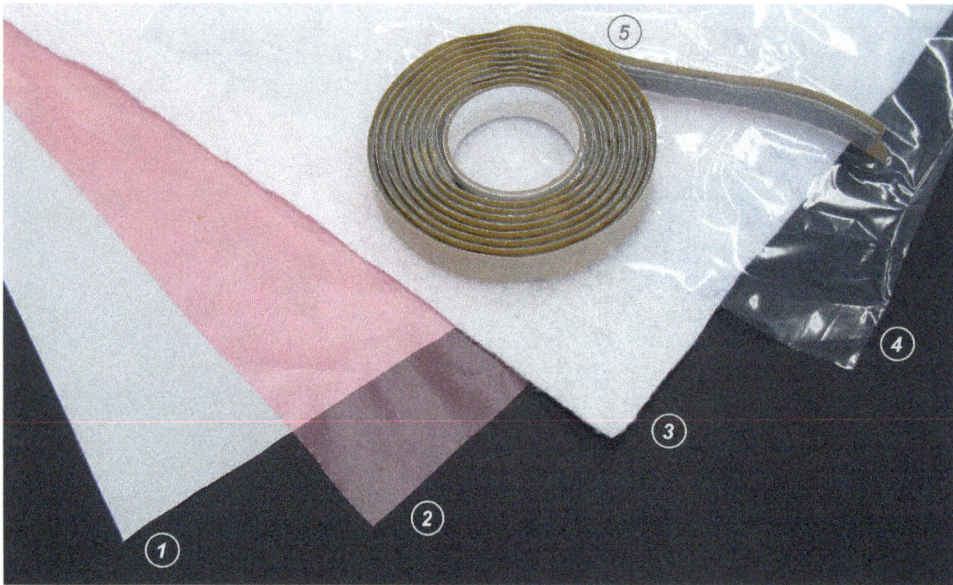

Some common materials used in vacuum bagging include:
1) Peel/release ply,
2) Release film
3) Bleeder/breather cloth
4) Bagging film
5) Sealant tape

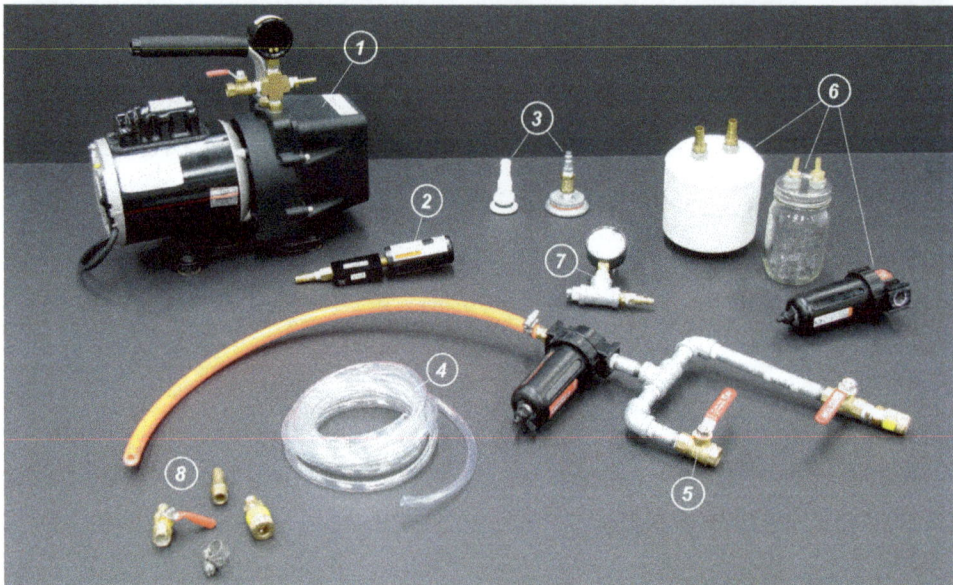

Some of the equipment used in vacuum bagging includes:
1) Vacuum pump
2) Vacuum generator
3) Vacuum connectors
4) Vacuum line/hose
5) Bleed-off and control valves
6) Resin traps
7) Vacuum gauge
8) Various air fittings.

Multiple bags can be evacuated at one time using a manifold system - like this shop air-powered vacuum generator setup that can facilitate four parts at once.

trated in one square foot of space! The amount of pressure that actually bears down on the laminate, though, is limited by the capability of the pump producing the vacuum, and the altitude at which the vacuum-bagging procedure is being done. Vacuum pumps come with varying capacities and maximum vacuum levels, and atmospheric pressure drops approximately 1/2 pound per square inch with every 1000 ft rise in elevation. The same 12" by 12" vacuum bagged panel at 6000 ft of elevation will be subject to about 11.7 psi of atmospheric pressure, or 1685 pounds of force - a 432 pound difference between sea level and the Rocky Mountains. Luckily, in practice, the average fabricator will probably not notice much difference between the quality of the parts produced at such widely varying altitudes.

Because of their potency in forming high-quality composite laminates, vacuum-bag molding techniques are used extensively in aerospace, military, and high-performance automotive applications. For these high-end applications, vacuum-bagging is often used in conjunction with "pre-preg" materials (which will be discussed in the next chapter) in high pressure autoclaves that substantially raise the atmospheric pressure around the laminate and mold for enhanced part quality. Considering the astronomical price of autoclave equipment (or even the limited access to it), the techniques shown in this chapter focus on vacuum-bagging with standard atmospheric pressure alone.

To perform vacuum-bagging correctly, a specific set of procedures must be followed. First, the mold is prepared with release agents and the laminate is laid up in the mold, as with a typical layup. Next, a series of vacuum-bagging materials (consumables in the form of specialty films and cloths) are applied over the laminate. These bagging materials are sealed around the mold's edges and a vacuum is then applied to the laminate. Once the laminate has cured, the bagging materials are removed and discarded. The result-

Vacuum-bagging Lamination Schedule

Resin System: Epoxy
2 Layers - 5.5 oz. Carbon (4HS Satin)
2 Layers - 8 oz. Fiberglass (Plain Weave)
PVC Foam Core
2 Layers - 19.7 oz. Carbon (12K - 2x2 Twill)
1 Layer - 5.5 oz. Carbon (4HS Satin)
Selective Reinforcement at Fasteners

Materials and Supplies

Bagging Film, Peel Ply, Release Ply
Sealant Tape and Masking Tape
Mold Release/Parting Wax
Colloidal Silica Thickener
Mixing Cups and Sticks
Paintbrushes and Disposable Gloves
Ruler and Utility Knife
File or Rasp
Clean Rags/Cloths
Scissors
Oven
Power Drill with Forstner Bits
Rotary Tool with Cut-off Wheels
Miscellaneous Fasteners/Hardware
Vacuum Connector
Vacuum Source and Vacuum Tubing

The vacuum bagging lamination schedule, along with materials and supplies, used in this demonstration.

Begin by applying wax/release to the mold.

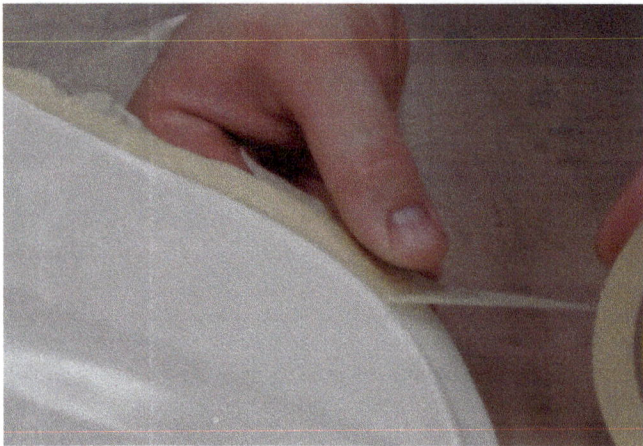

Tape off the flanges to aid in flange cleanup later.

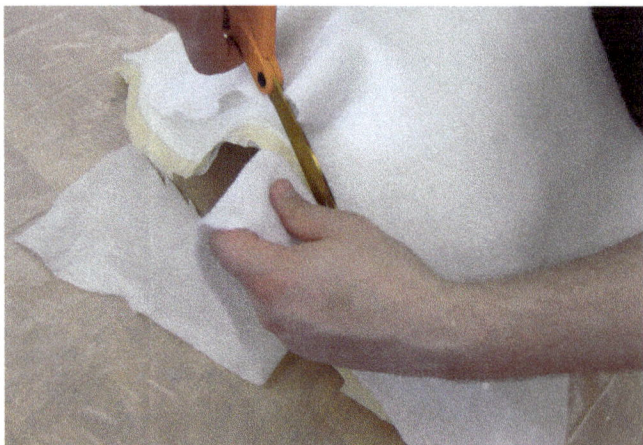

To make a cutting pattern for the reinforcement fabrics, bleeder/breather cloth can be put over the mold and then cut to size.

ing high-quality laminate is then demolded, trimmed, and finished prior to being put into service.

Peel/release ply is the first layer applied over a composite laminate when secondarily bondable surfaces are needed. It is composed of strong cloth (such as nylon or polyester) coated with release so it can be pulled away from the laminate after cure. This ply creates a rough surface on the laminate while letting resin flow through it. The fine pores in the peel ply's weave promote transfer of resin from the laminate to other bagging layers so excess resin can be removed. In doing so, these pores also form a uniform texture on the surface of the laminate that aids in later secondary bonding of components or additional laminate plies.

Release film (also called "perforated film") is a thin film of polyethylene, nylon, or Teflon perforated with a pattern of small holes. These holes allow resin to flow away from the laminate (though less freely than with peel/release plies) while still preventing bonding between the bleeder/breather cloth and the laminate. Though it may look somewhat similar to peel/release ply, release films are typically less expensive, but can also tear easily. They also do not produce the rough, bondable surface of a peel ply. However, without this protective release film layer, resin from the laminate will form a tenacious bond with the bleeder/breather cloth used in the vacuum bagging process - something that can only be remedied through hours of grinding and sanding to remove the cloth, or the unfortunate trashing of an entire composite part.

In theory, a bagging system can work well with just peel/release ply and bleeder/breather layers alone, but in practice, demolding is much easier when release film is between them, or else when two layers of peel/release ply are placed atop each other. This is especially true of parts that have significant curvature or shape to them. Without adequate release materials between the laminate and the bleeder/breather cloth, the rigidity of the resin-saturated bleeder/breather cloth can make demolding and separation of the bagging materials very difficult.

Bleeder/breather cloth is a non-woven cloth made of high-fill polyester that is similar in appearance to quilt batting, but without the chemical binders that would interact with resin and disrupt airflow like actual quilt batting. This layer is used to wick up excess resin (the "bleeder" function) and transport air out of the laminate to the vacuum pump (the "breather" function). Originally, these two functions were performed by different layers of material, but current bleeder/breather cloths perform these tasks as one combined cloth. Bleeder/breather cloth is so efficient at performing its job that it can actually rob the laminate of too much resin if full vacuum is initially applied to a wet layup with very thin resins. To prevent this, a pressure bleed-off valve can be added in the vacuum lines (as described below) to reduce the vacuum until the resin begins to gel. Bleeder/breather cloths are available in different material weights, depending on the pressures that will be exerted on the vacuum bag. Typical four or five ounce bleeder/breather cloth will work well for atmospheric air pressure applications - such as home or small shops that don't employ an autoclave.

Bagging film is made from flexible polyethylene (for low temperature bagging) or nylon (for high temperature oven or autoclave use). It is used to completely seal off air from the open face of the laminate and to compress the bagging materials against the composite with the atmospheric pressure acting on it. Take care to avoid cutting or poking holes in the bagging film during handling - air leaks in bagging film can be notoriously difficult to locate but can severely compromise the vacuum within the bag. If relatively flat or simple shapes are being bagged, inexpensive 6-mil (.006" thick) polyethylene film, commonly employed as a painting drop cloth, can be used. Likewise, such forms can also be bagged using special bagging

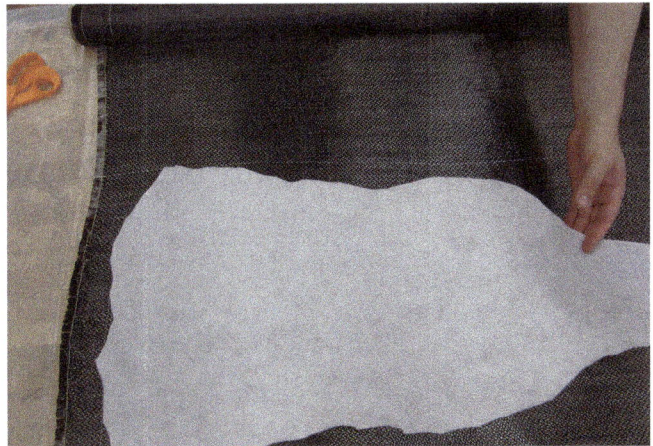

Trace this pattern and use it to cut reinforcements to the correct size.

Cut out all the needed layers for the layup prior to mixing any resin—working time is critical with vacuum-bagged layups.

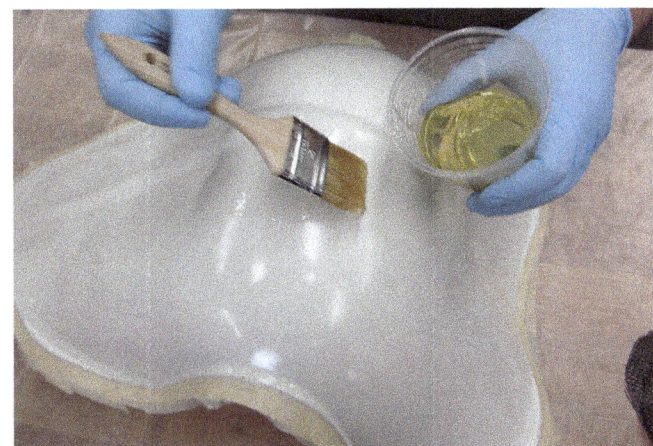

Spread a thin coat of resin over the mold to saturate the reinforcements from beneath...

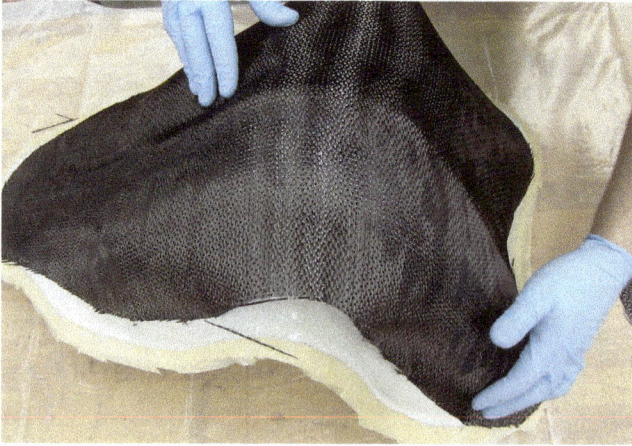

...and lay on the first layer of reinforcement.

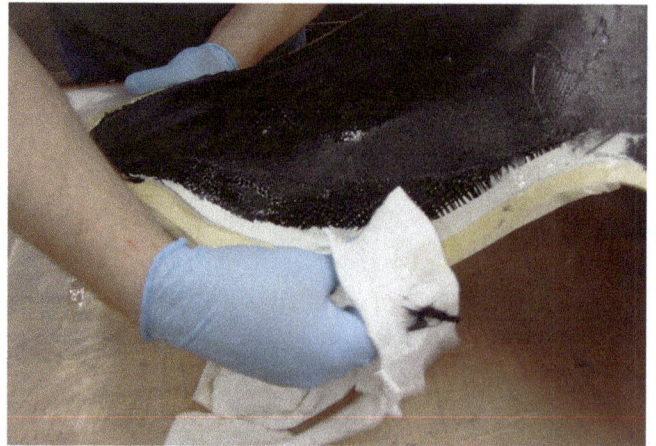

Occasionally wipe up any resin that may spill onto the taped flange.

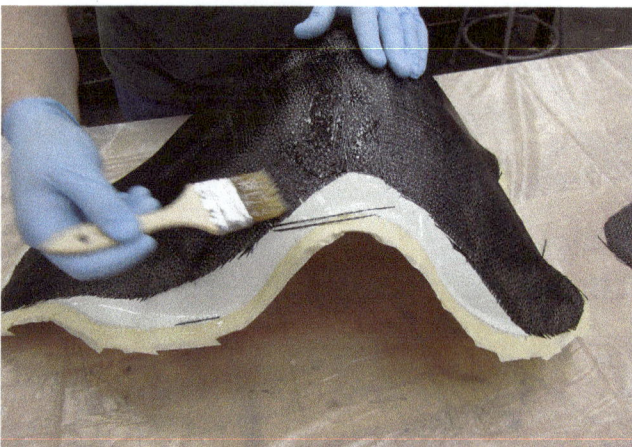

Apply additional resin, as needed, to completely impregnate the fabric, but avoid using more resin than is really necessary.

After the layup is complete, drape a layer of peel/release ply over the laminate to make a secondarily bondable surface.

Trim away unneeded edge material throughout the layup.

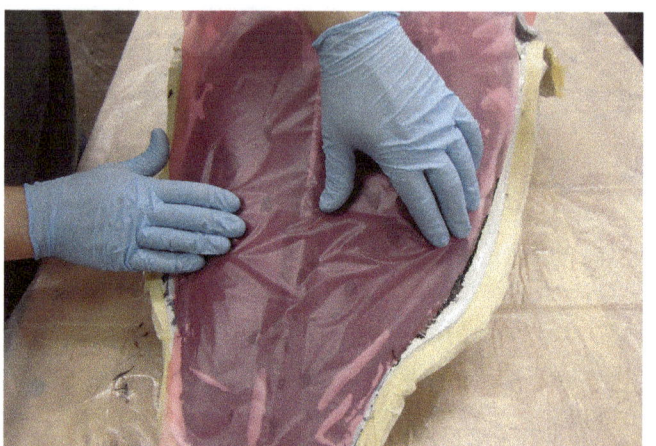

Next, apply a layer of release film...

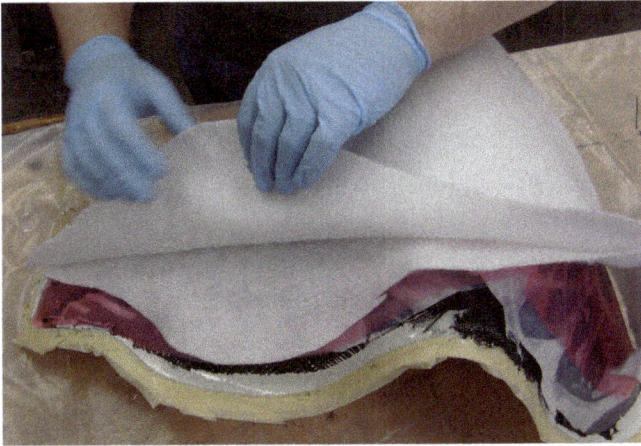

...and then bleeder/breather cloth. If the layup is very "juicy", apply multiple layers to wick up the excess resin. Bleeder/breather will only be effective if it will not become completely saturated.

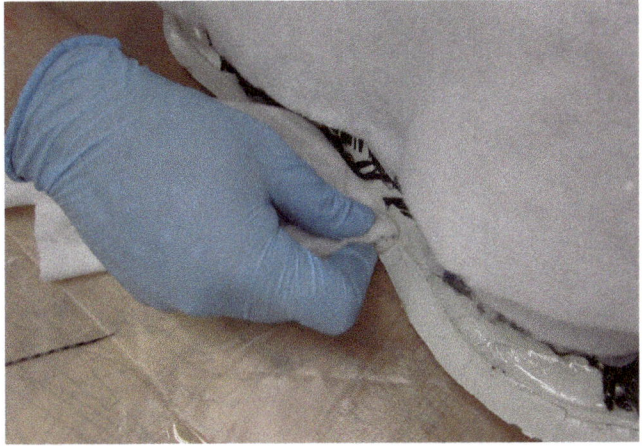

...and use acetone and a rag to remove any resin from the flange—sealant tape will not seal to resin-covered surfaces.

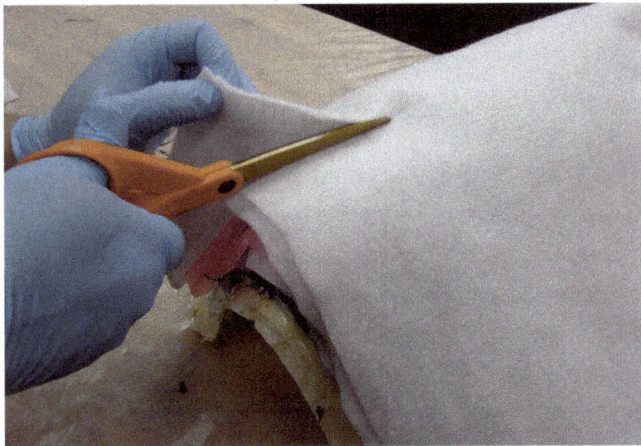

Cut reliefs into the fabric if needed to help it lay down on the part without making too many wrinkles.

Apply sealant tape to the entire flange...

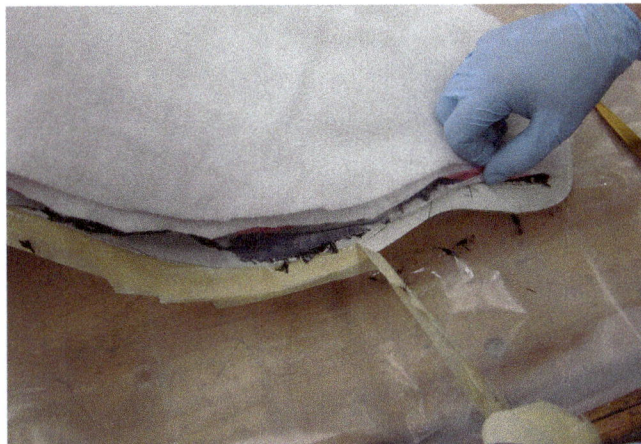

Remove the masking tape from the flange...

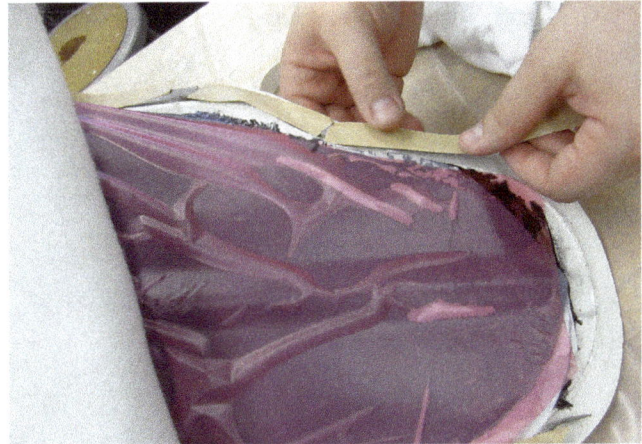

...pressing down with firm pressure to create a good bond.

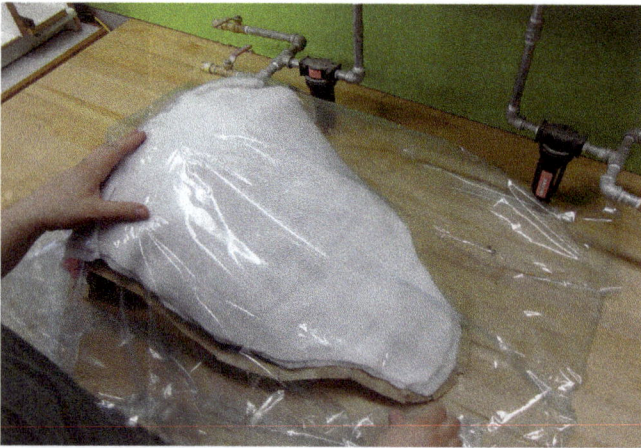

Drape bagging film over the entire part, leaving enough material at the edges to accommodate the contours of the part.

Peel the backing off the sealant tape...

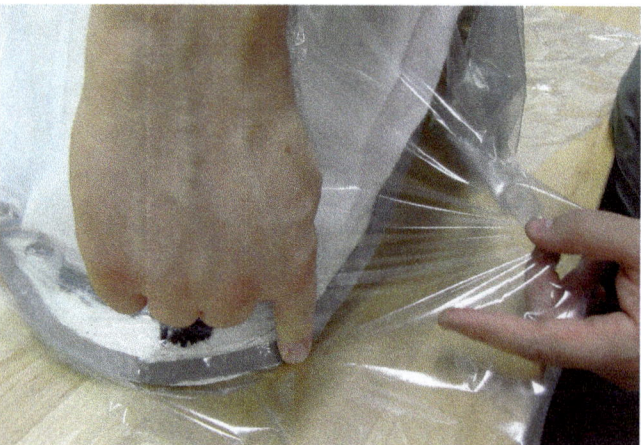

...and press the bagging film onto the tape—at first with light pressure until the whole part is enclosed, and then firmly to complete the seal.

"envelopes" that come as large flattened tubes of polyethylene plastic that only need to be sealed at either end. For bagging parts with complex shapes and curves, more flexible or stretchy bagging film is recommended.

To close off the edges of the vacuum bag, use a high tack silicone sealant tape. This specialty tape looks similar to weather stripping, but maintains its tackiness indefinitely, and to very high temperatures. Sealant tape can be stretched and formed to complex mold and bag shapes very well. It is typically available in rolls that must be stored flat on their sides to avoid slumping and sticking of the silicone onto itself - a property of this tape that can quickly ruin poorly stored rolls. Avoid getting dirt, debris, or resin on the sealant tape as these will lessen its ability to create a good vacuum seal.

Vacuum-bagging techniques can be used on both wet and dry layups, with and without sandwich cores. However, since wet layups are typically much messier than dry layups (like those that use pre-preg or resin transfer molding), special care should be taken to contain and control where the liquid resin goes during layup so it will not get on the mold flanges and compromise the vacuum seal. This can be done by taking precautions during layup (as shown in this chapter's demonstration) to cover up and protect the mold flanges. Additional bleeder/breather cloth may also be needed with extra juicy wet layups to wick away excessive amounts of resin.

Vacuum-bagging works very well for a variety of part designs, but its eventual use must be designed into the mold for best performance. This includes creating an air-tight mold that is free from pores or cracks through which air might leak. It also requires designing flanges into a mold onto which a vacuum sealant tape can be applied to close off the mold and bagging materials. Some specialized production molds even contain locking mechanisms that seal off the edges of a rubber or silicone diaphragm over the laminate in place of traditional, more time intensive bagging materials.

Even though vacuum-bagged composites are laid up using the same methods and equipment as those created in open molds, specialized shop equipment is required to perform the vacuum-bagging procedures. This equipment includes the items explained below, each of which is available from a variety of composites suppliers.

A vacuum pump (usually driven by an electric motor) creates a high vacuum by using a piston, diaphragm, or vanes to pull air from the vacuum bag. These pumps are essentially air compressors that are used in reverse, but they have seals specifically designed for continual high-vacuum use. Vacuum pumps are very different from the typical home or shop vacuum that is used to clean up messes; they generally produce about three to five times the vacuum (though at less volume than a vacuum cleaner) and can run under high-vacuum for long periods of time without problem. Care should be taken to make sure that they are well-maintained (especially with oil-based pumps) and that nothing other than air makes its way into the pump's vacuum inlet.

Composite suppliers sell a variety of pump sizes to meet fabricators' needs, the recommendations for which are based upon the surface area of the parts you intend to mold.

If you already have an air compressor in your shop that can push air at a couple cubic feet per minute (CFM) or more, a cost effective alternative to a vacuum pump may be a vacuum generator. This is a device that is simply attached to an air line yet produces a very high vacuum. As high pressure air from a compressor moves over an internal orifice in the vacuum generator, it pulls air out from the vacuum bag. Vacuum generators cost a fraction of what a vacuum pump does, yet they have no moving parts themselves and require no maintenance.

Vacuum connectors join the vacuum bag to the vacuum lines and pump system. High-end

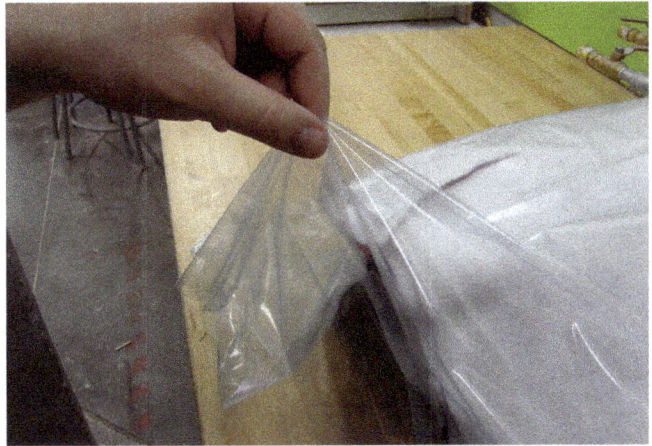

For areas with complex contours, leave enough extra film to make a "pleat" or "tuck". This will allow the bagging film to conform to the mold without tearing.

Seal up the pleat with extra sealant tape.

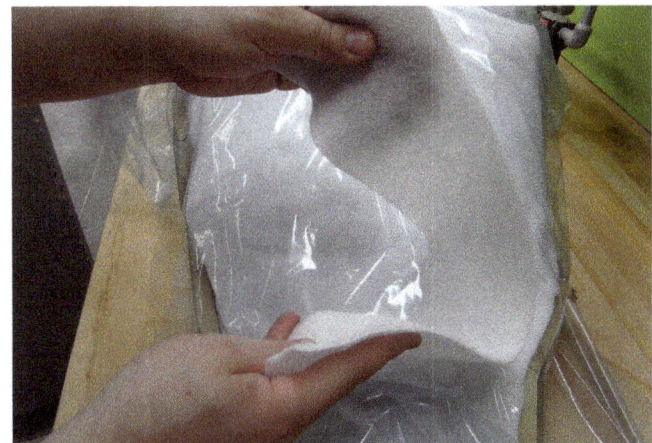

Take some extra bleeder/breather cloth...

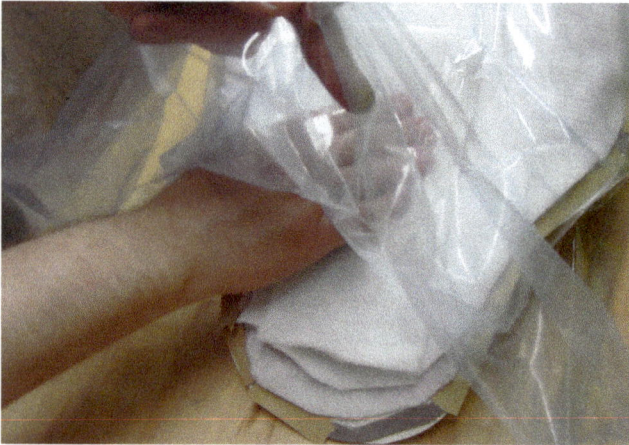

...and insert it where the vacuum connector or vacuum line will be placed on the part.

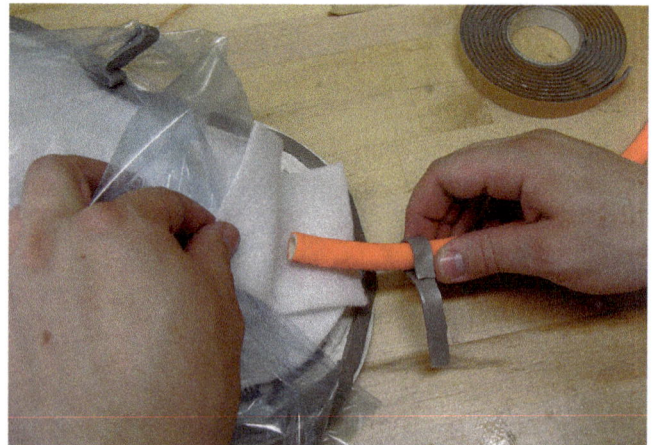

...insert it between layers of bleeder/breather cloth...

To create a quick vacuum connection using a vacuum line, take a piece of sealant tape...

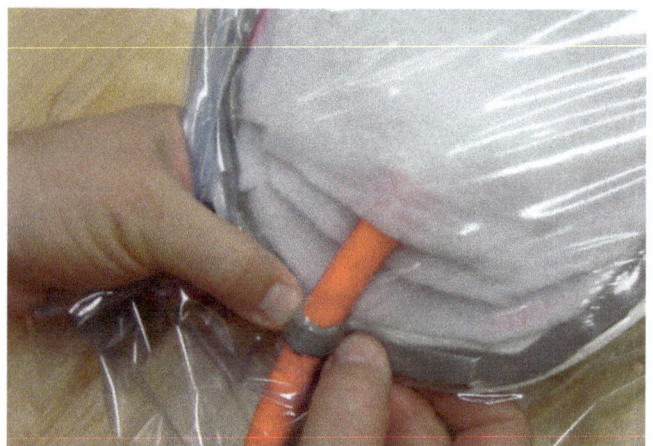

...and seal it to the edge of the flange.

...wrap it around the vacuum line...

Through-bag connectors can be inserted by cutting a small slit in the bag...

...placing the bottom of the connector in the bag...

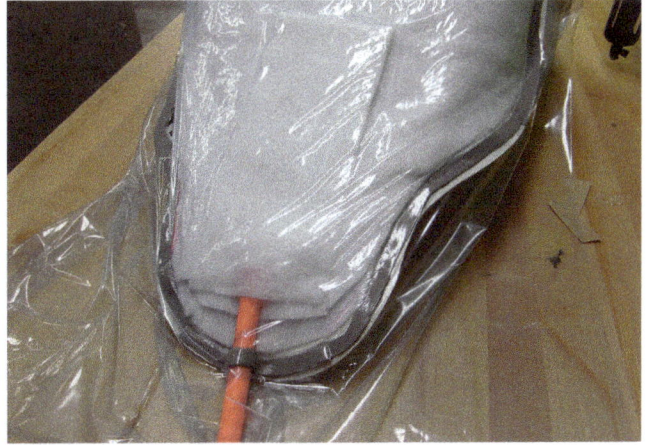

Double check the bag's seal prior to applying vacuum...

...attaching the top of the connector through the slit and twisting it onto the connector's bottom piece...

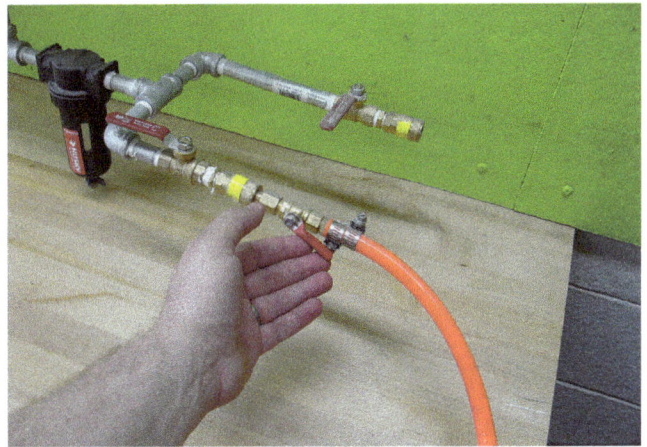

...and then slowly apply the vacuum...

...and then sealing the bag and connecting it to the vacuum line.

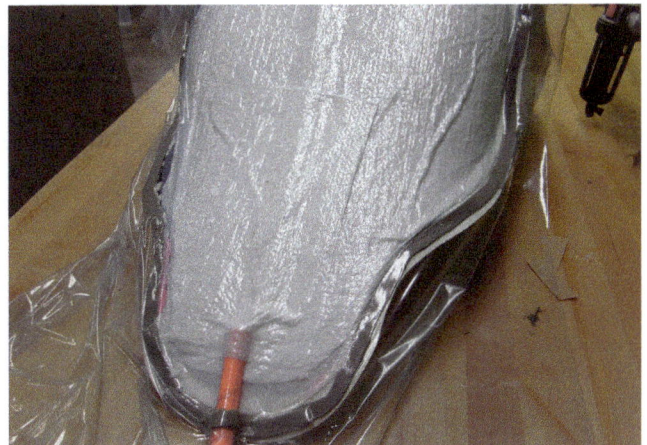

...until the part is completely evacuated. Check for any possible leaks.

If you notice any large wrinkles, relieve the vacuum with the bleed-off valve and smooth out the bag as the vacuum is applied again.

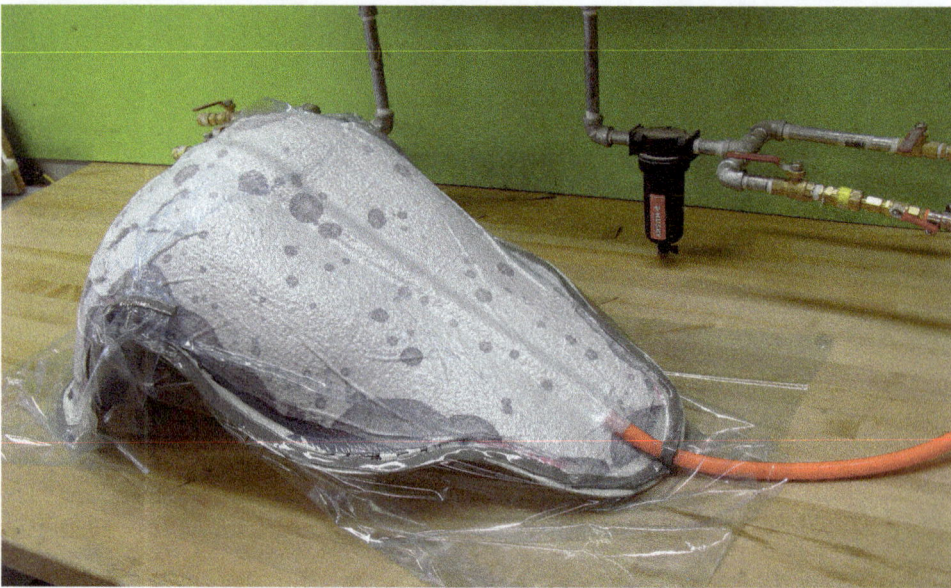

Re-check for any vacuum leaks, and then allow the part to cure.

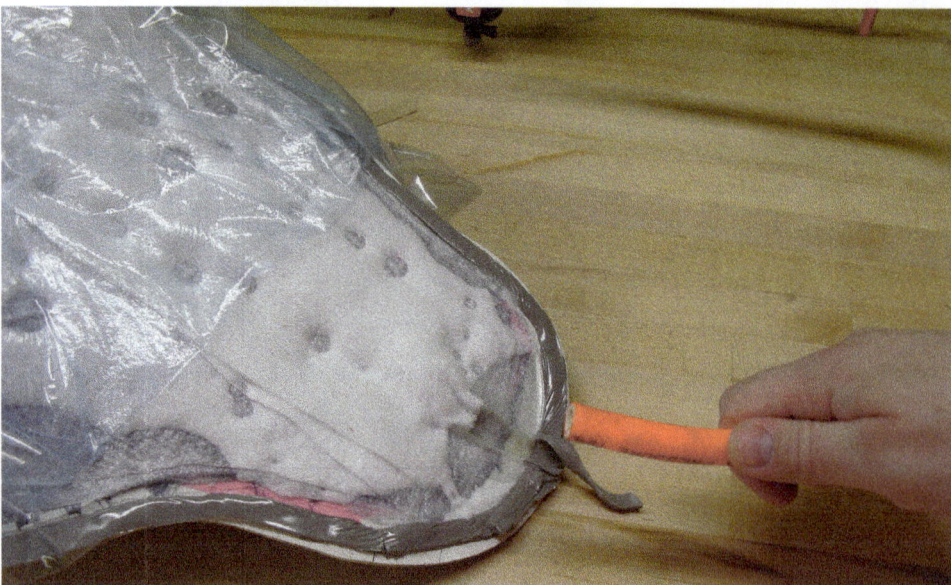

After cure, remove the vacuum line/connector...

connectors contain a valve and an air fitting that help ensure a good vacuum after the vacuum line has been removed. Less expensive vacuum connectors will still seal well, but will require the vacuum to be connected to the vacuum-bagged part full time unless an adequate shut-off valve is included in line with the connector. For less sophisticated systems, a simple vacuum hose sealed at the vacuum bag's edge can work equally well in place of a vacuum connector.

Vacuum lines are made of non-porous extruded nylon, vinyl, or rigid rubber tubing. They can be cut to any needed length and connected with common air fittings. When selecting tubing for vacuum lines, chose one that has sufficient wall thickness (to prevent crushing under ambient air pressure) but that is flexible enough to be managed easily during the bagging and curing processes.

When multiple bags are being evacuated at the same time, a manifold or system of connectors and valves can be used to control the vacuum to each bag. Inexpensive air fittings from the local hardware store can be used to create such a manifold, but thread sealant, hose clamps, or Teflon tape should be used to guarantee a tight seal at each joint.

To control the amount of vacuum pulled from a bagging system, a bleed-off valve is placed inline with the vacuum bag. A vacuum bag that evacuates too quickly may cause problems for laminates with thin resins or complex geometries. If a resin is too thin, an instant full vacuum may simply draw all the resin from it into the bleeder/breather cloth, starving the laminate and creating a weak composite. Likewise, a quickly applied vacuum can also cause the vacuum bag to constrict too quickly, preventing the bagging film from forming adequately into cavities and corners. The bleed-off valve allows air to enter the vacuum system to control how much (and how quickly) vacuum is applied to the bagging system.

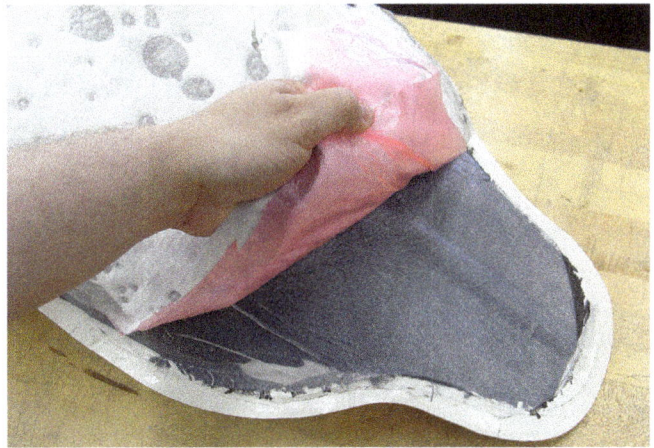

...and peel off the bagging materials.

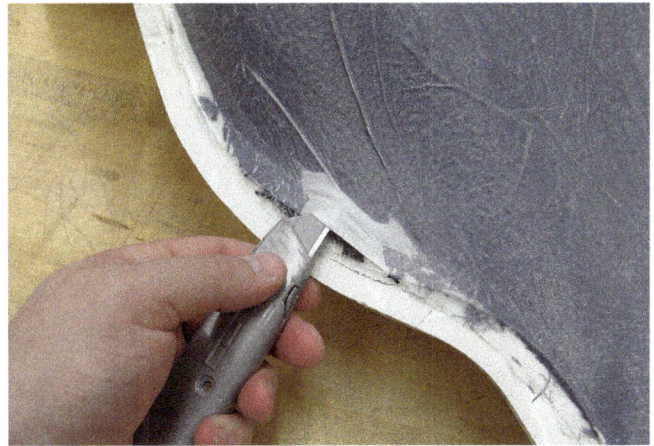

Carefully use a utility knife to peel back any stubborn materials that may have adhered themselves to the mold.

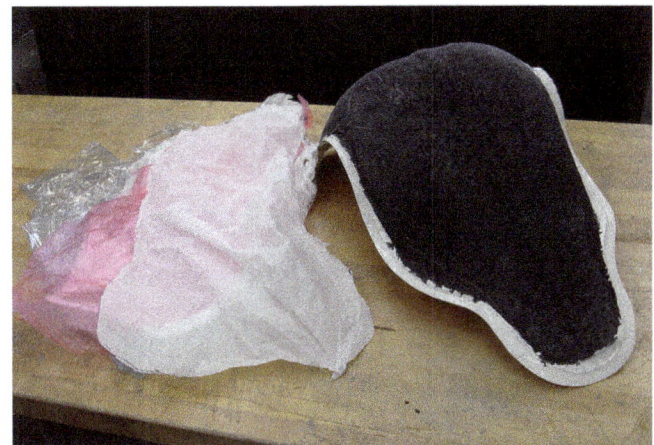

Used bagging materials can be discarded.

To demold the part, insert a wedge under the part in several spots...

...and pry it off the mold.

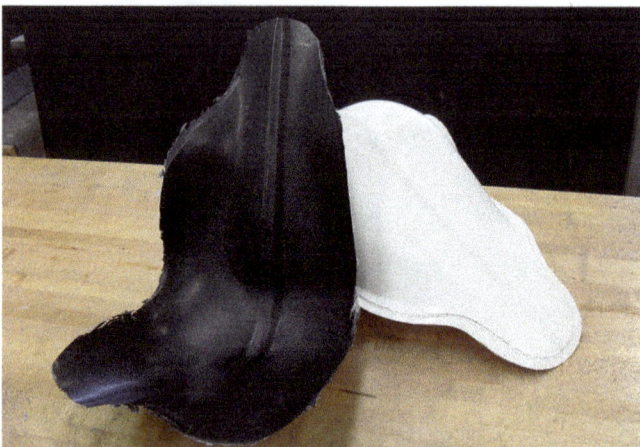

The bagged part can now be trimmed if no additional bonding processes are required.

A resin trap is a small container that is affixed to the vacuum line as a safety measure to protect the pump from being infiltrated with resin - a serious problem that can quickly turn an expensive pump into an anchor. With this simple device, any resin that gets inadvertently pulled into the vacuum line will drip into the resin trap rather than flow into the vacuum pump. A resin trap is especially important when vacuum-bagging extra juicy wet layups or when performing resin transfer processes - the later of which may require an especially large resin trap.

Resin traps can be constructed of practically any container that is air-tight and can accept fittings for connection to the vacuum lines. Large resin traps can be purchased from composites suppliers, but other equally effective solutions include a Mason jar with fittings in the lid, sealed PVC pipe sections configured in a "T", or even line dryers for pneumatic systems.

To monitor the vacuum being achieved in the vacuum system, a vacuum gauge is indispensable. Vacuum gauges measure the vacuum of a system in inches of mercury ("Hg) and can be quickly connected to the vacuum line or pump through a Y or T-connector after the resin trap.

If a vacuum-bagged laminate needs to be smooth on both sides, a caul plate (a smooth, removable form made from sheet metal or other non-porous material) can be applied directly to the open face of the laminate, covered with bleeder/breather cloth and then bagged. However, the caul plate may hinder excess resin from migrating out of the laminate to produce a truly optimized composite - though the aesthetics of the part will be markedly improved where the caul plate touches the laminate.

STEPS FOR BASIC VACUUM BAGGING

To demonstrate common vacuum-bag molding techniques, this chapter will show the layup of a go-cart sport seat over a male mold, followed by the application of all bagging materials. A vacuum will then be drawn and the part

To ease secondary bonding, use a file to remove any wrinkles that may have been left on the laminate by the bagging process.

Use water to clean off any dust or debris prior to bonding.

A PVC foam core is used to create a sandwich structure on this compound curved part. To apply such a core, cut the foam to size with a knife...

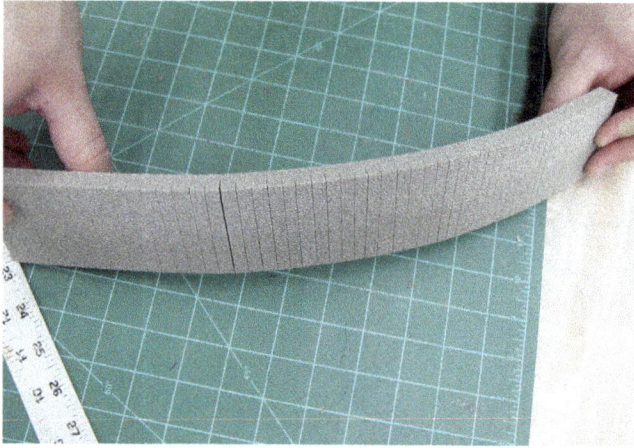

...and score it (with shallow slits that are closely and evenly spaced) so it can be readily bent.

Finish any other needed shaping with a knife.

A heat gun can loosen the PVC foam a bit so it will better conform to the part.

The properly processed foam pieces should hold their shape relatively well.

Double-check the foam for correct fit. Small gaps can be filled later with resin paste.

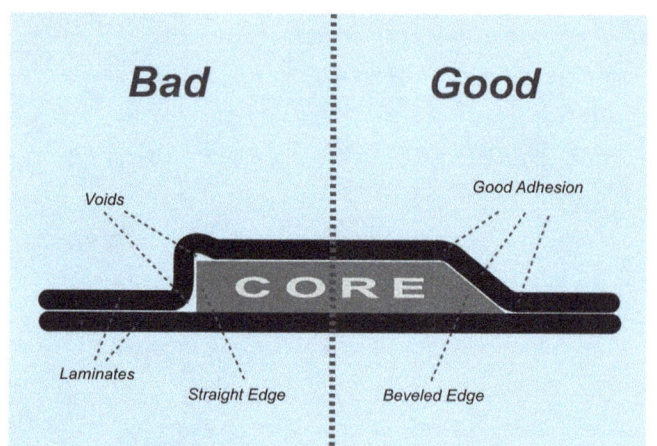

A properly applied core should have beveled edges to promote adhesion between the top and bottom plies in the sandwich construction.

Use a knife to cut a bevel at the foam's edge…

Compound curved foam sections will need additional heating to be shaped correctly.

…sand out any waves or large imperfections…

An oven at about 350 degrees Fahrenheit will soften the foam well.

…and then use shop air to blow away any foam dust.

The heated foam will discolor a bit, but can be easily formed with a little pressure.

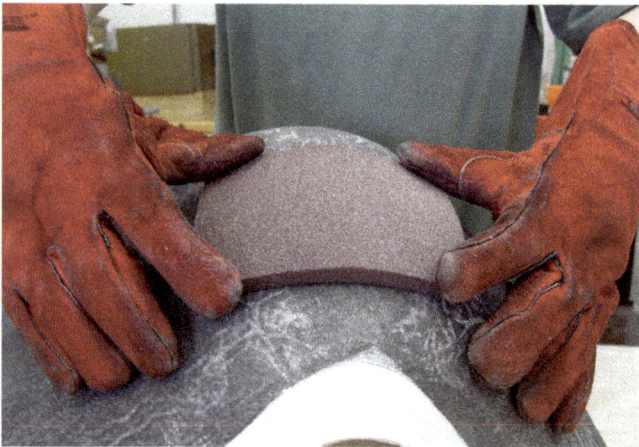

Press the heated foam over the surface it will be bonded to…

…and then trim it to size and add the needed edge bevels.

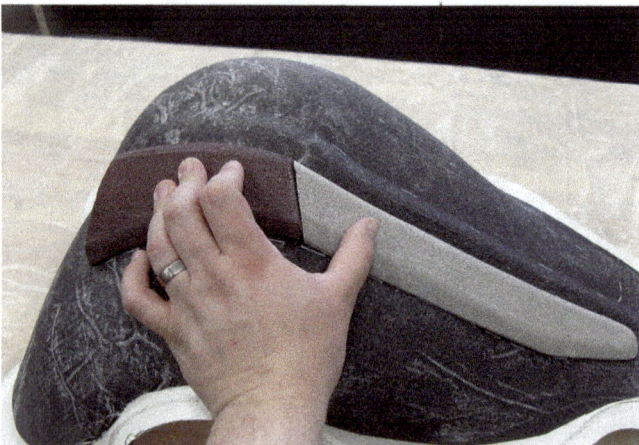

Double-check the pieces for fit, and leave alignment marks on the laminate to help during layup.

will be allowed to cure. To further rigidify the seat, a foam core will be added to the cured seat, and then laid over with additional composites, bagged, and cured a second time. It will then be demolded and prepared for trimming and finishing.

Collect all materials and tools prior to beginning the layup, and determine a plan for the entire molding and vacuum-bagging process beforehand so everything will run smoothly. Prepare the surface of the mold with wax release, and then apply masking tape to all the mold flanges. Apply a coat of PVA liquid over the mold for extra protection. Use scissors to cut the reinforcement fabrics and then stack the reinforcement plies as listed in the lamination schedule so they will be easier to place in order in the mold.

Mix the resin and hardener/catalyst as recommended by the manufacturer. As always,

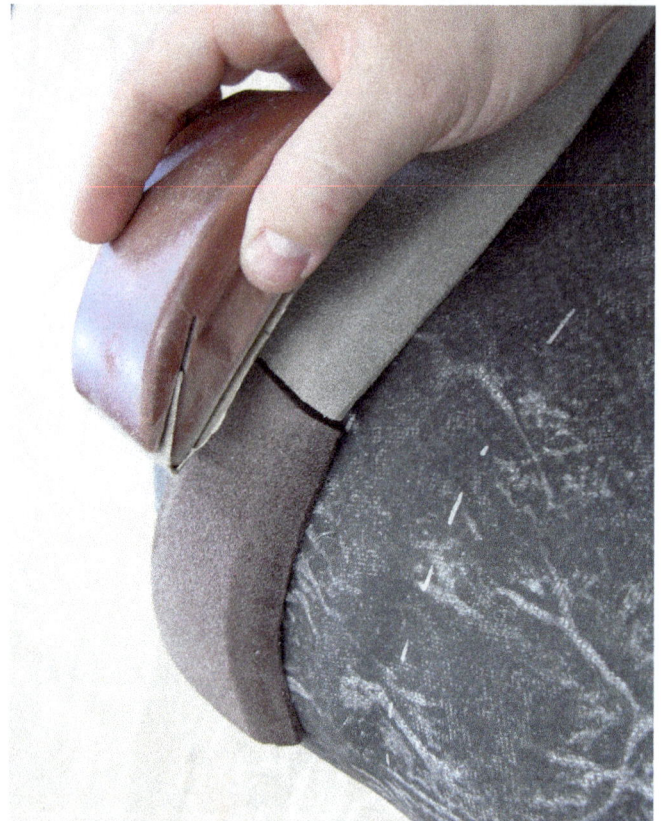

Sand any lips that occur between foam pieces—poor transitions between core pieces can weaken the sandwich laminate.

Mark and trim any additional pieces in preparation for adhesion.

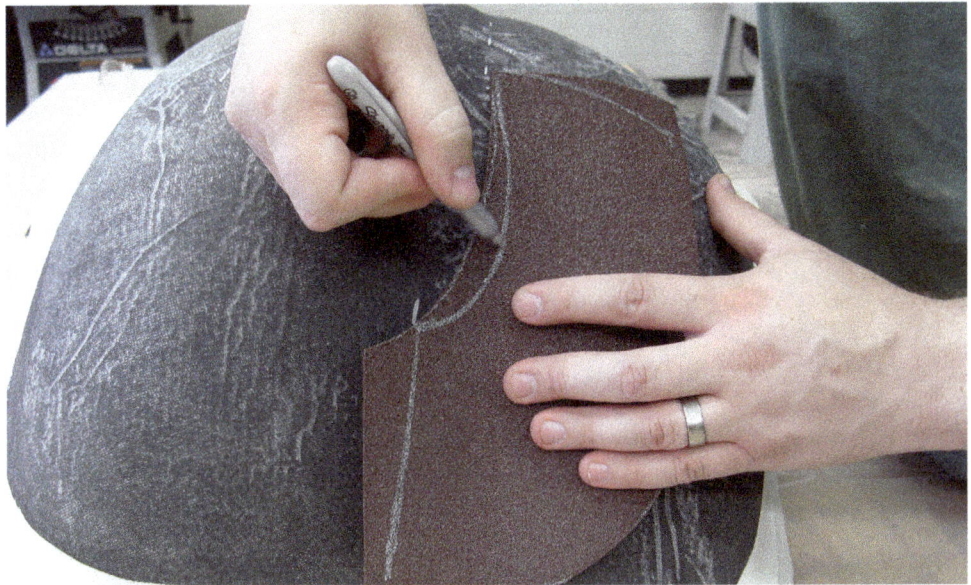

Create a resin paste by adding colloidal silica to the resin…

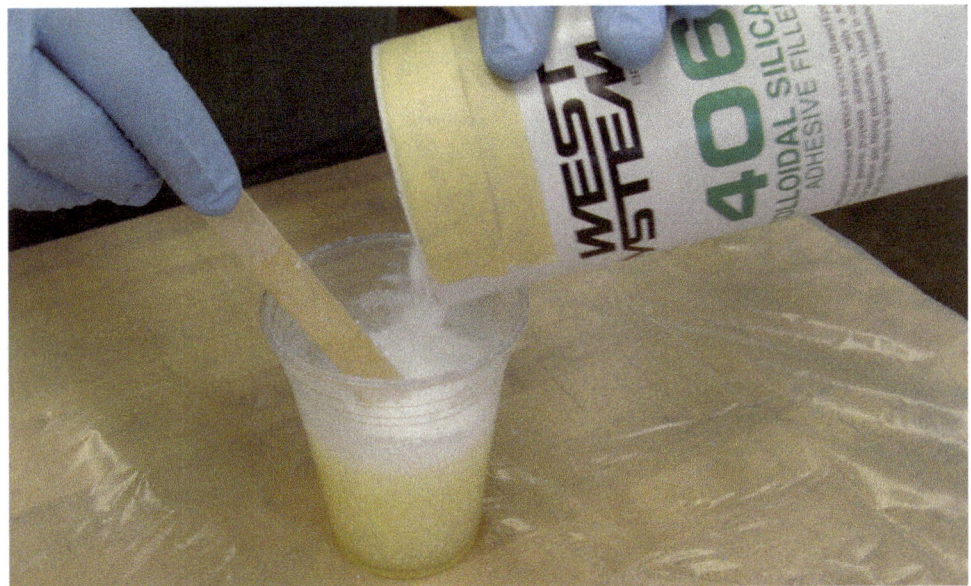

…and then spread it onto the core pieces.

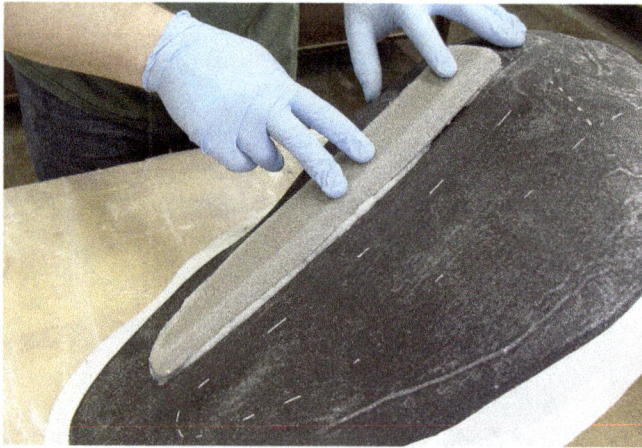

Press the core onto the laminate so that excess resin paste will be pressed out the edges.

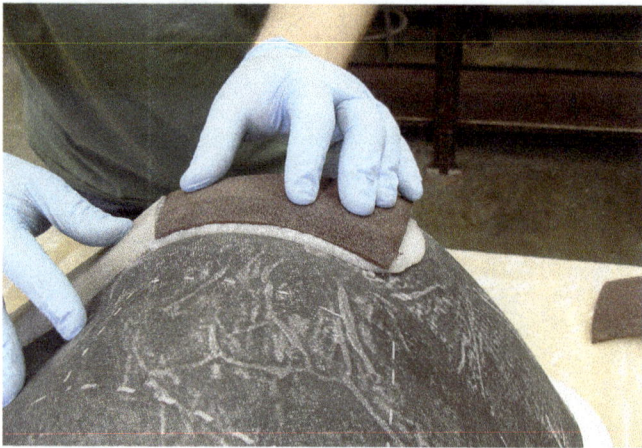

Add the remaining core pieces in this same way.

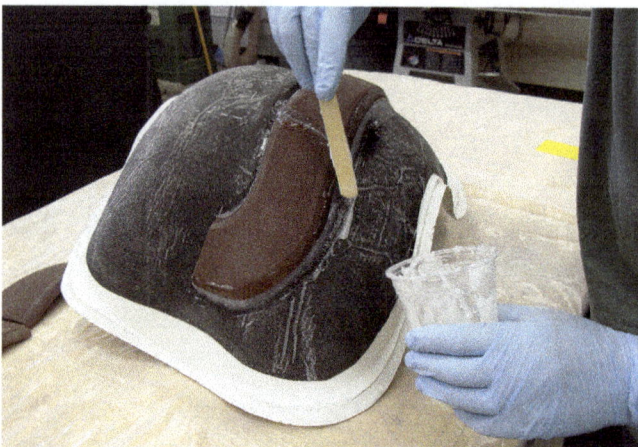

Use a stick (or your finger) to round over and remove any excess resin paste.

make sure to scrape the sides and bottom of the cup as much as possible during mixing so the resin and hardener will be completely combined. Mix only enough resin to apply to the laminate before it can begin to cure - and keep in mind the time difference between the resin's pot life and working time. With vacuum bagging, your timeframe to completion will be determined by the cure time of the resin because the resin needs remain liquid until the vacuum is applied, otherwise the process is pointless. Therefore, very small parts may still be laminated with fast cure resins without problem, but most large parts will likely require slow-cure resins.

Completely cover all the molding surfaces with resin to allow for better wet-out of the fabric from beneath while holding it in place. Carefully place the first layer of reinforcement fabric onto the mold, avoiding any tugging of the fabric that may warp the fibers. Apply light pressure on the fabric with a brush (for curved parts) or spreader (for flat parts) to help work the resin through the fiber. Add the remaining layers one at a time, and continue to add resin to the top of each layer, working it through as needed for full resin saturation and to remove bubbles and large voids from the laminate. Be sure that laminated materials are laying tight against the mold.

If adding a core to the layup to produce a sandwich structure, it is important to ensure thorough bonding between it and the laminate plies on either side of it. To do this, use thickened resin to fill in surface pores and provide a positive bond with the laminate (as shown in the previous chapter on compression molding). With wet layups, open-celled cores (such as honeycomb) may have problems with the cells filling with excess resin. To avoid this, honeycomb cores may be secondarily bonded to a roughened laminate surface that has previously cured with a peel-ply release layer over it or otherwise roughened up with heavy-grit sandpaper. Additional plies can then be laminated over the bonded

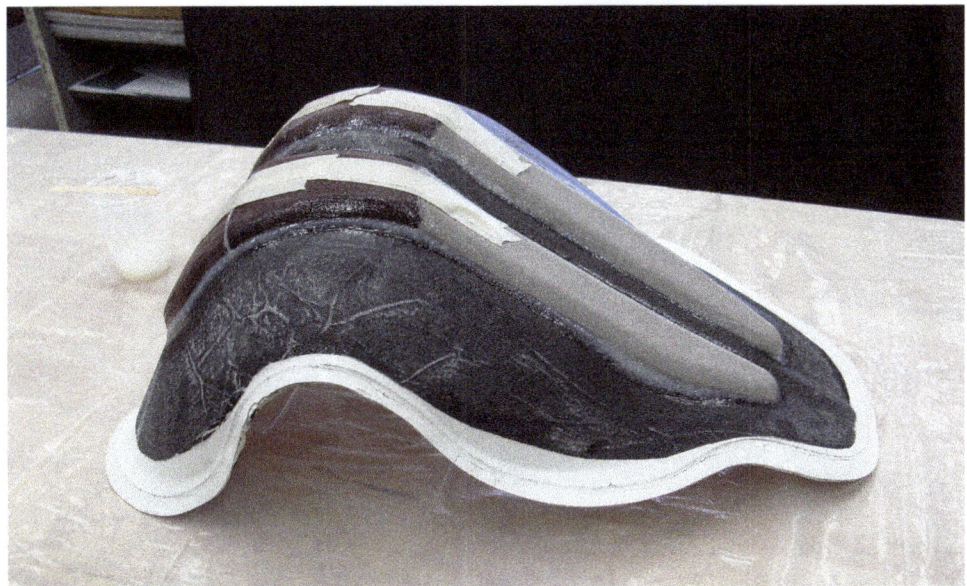

Masking tape can be used to keep the core pieces in place...

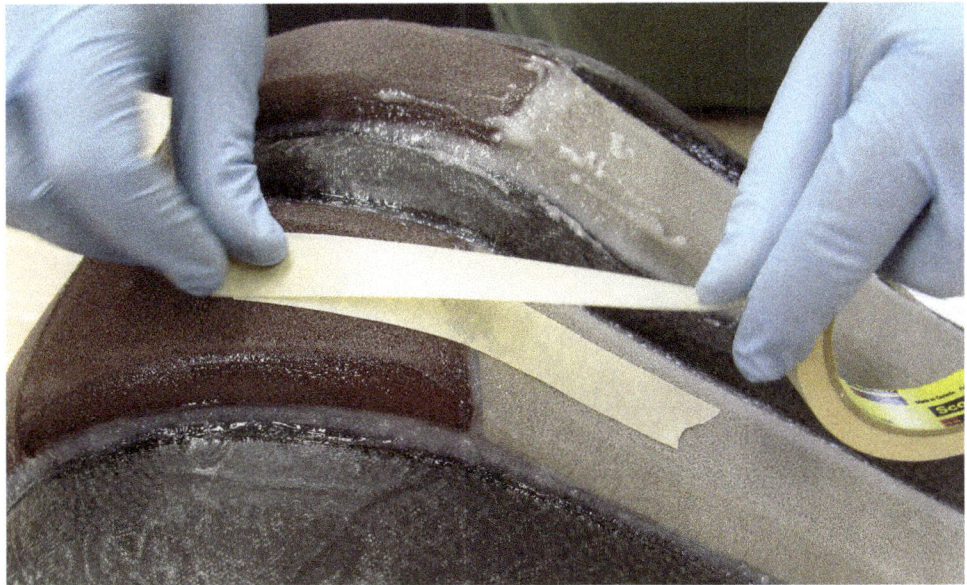

...until the resin begins to tack up.

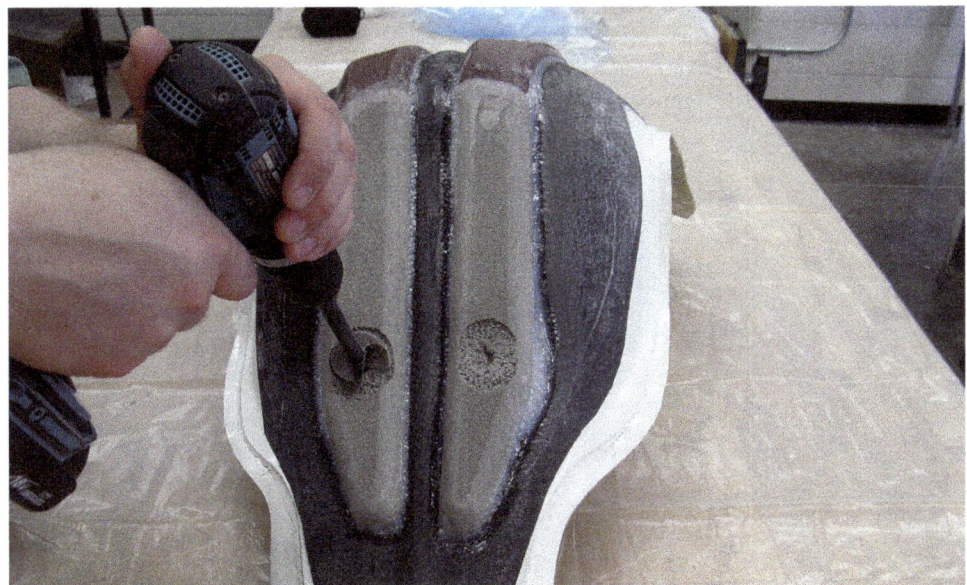

To apply screw fasteners into the lamination, use a drill with a Forstner bit to create a slight recess feature in the core where these fasteners will be.

core, re-bagged and vacuumed (as shown in this chapter's demonstration with a foam core). Cores in flat part sections can usually be laid up without problem , but this two-step core adhesion method is especially helpful when creating a sandwich structure from laminates that have compound curved or irregular surfaces. As shown, PVC foam can be pre-formed (cut, sanded, scored, or heat formed) to fit the surface of the laminate, and then resin paste added to fill in any irregularities between the laminate and core. This method of secondarily bonding the core to the laminate also helps minimize movement in the reinforcement weave of aesthetic surfaces - a problem that can arise when the core moves and pushes against the laminate as vacuum pressure rises.

Properly shape the core edges so the composite plies above and below the core will bond well to it and have minimum distortion or voids at the core interface. To facilitate this, it is best to bevel the edge of the core (a step called "scarfing") so the transition between the plies will be less disrupted and less apt to form wrinkles and cavities near the core.

Once all the laminations have been placed in the mold, apply the bagging materials in this order: Peel/release ply (if necessary), release film, and bleeder/breather cloth. After these layers are in place, carefully remove the masking tape from the mold flange, and in its place apply silicone tape to all the flanged edges of the mold. Apply an extra strip of breather/bleeder cloth where the

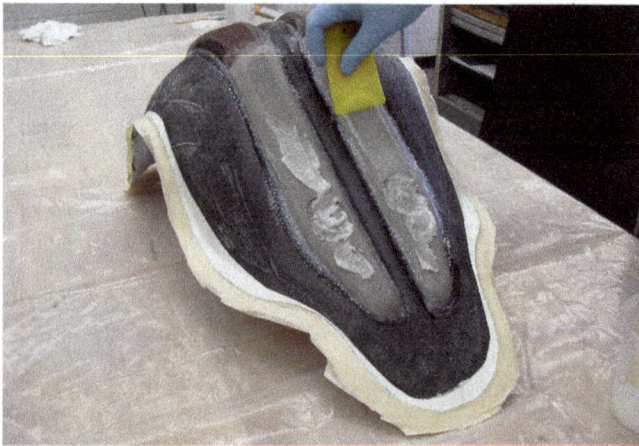

Apply resin paste over the top of the entire core...

...and then apply small patches of reinforcement and resin to the foam recesses. These will reinforce the fasteners from below and prevent crushing or failure of the core.

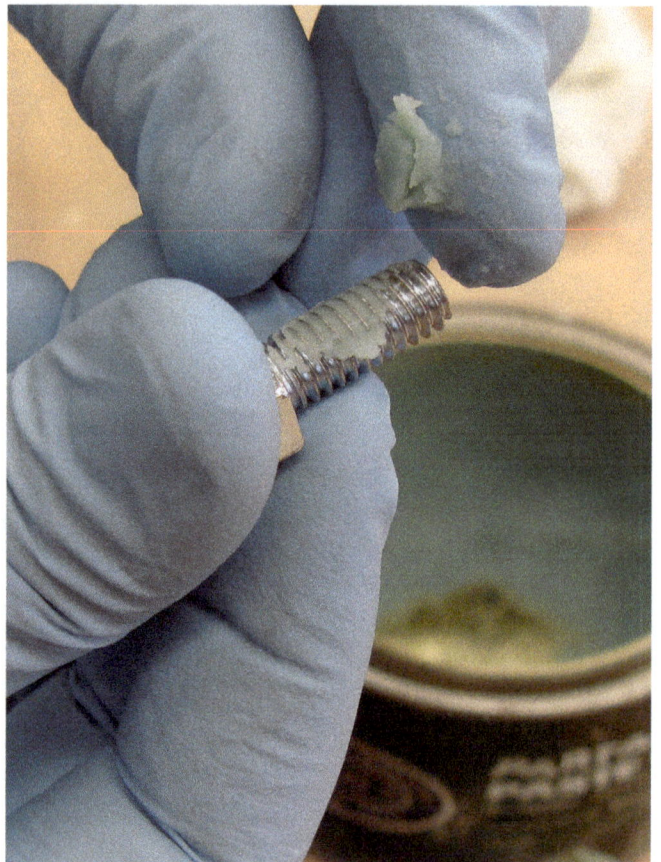

T-nuts will be used as the fasteners for this project, but they can fill with resin unless plugged. Waxed bolts can help in this regard.

*With the waxed bolt
threaded into the T-nut…*

*…the T-nut can be placed
over the reinforcement
patches.*

*To hold the T-nuts in place,
spread out the fibers from
an additional patch…*

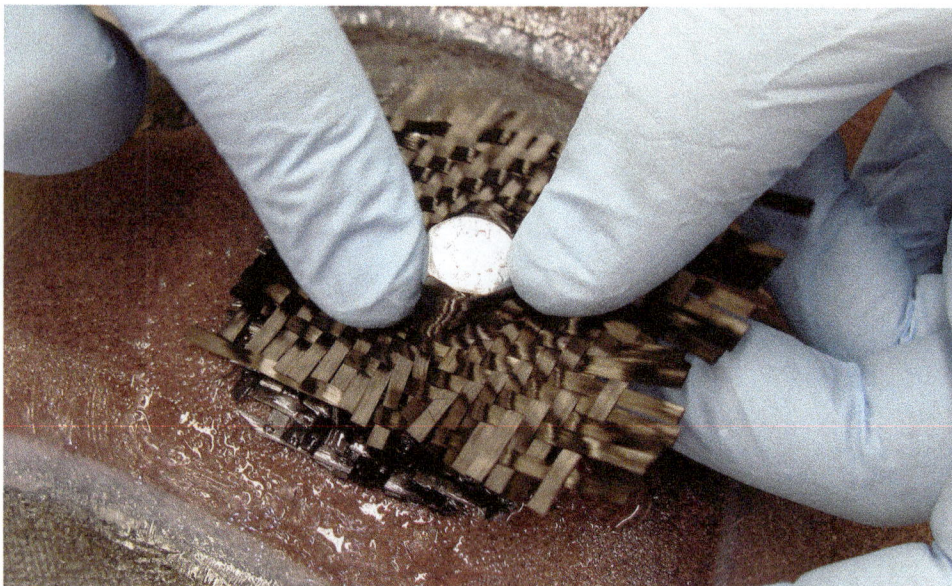

...and place it over the flanges of the T-nut.

Wet out the patch and proceed with the rest of the layup.

Layup additional plies of reinforcement material, as outlined in the lamination schedule.

vacuum connector will be positioned through the vacuum bag. This extra cloth will serve as an air evacuation pathway to the vacuum line while providing extra padding to keep the vacuum connector from leaving an imprint on the part during cure. Apply the bagging film to the top of the mold, sealing it to the silicone tape all around the mold. Create "pleats" and "tucks" as needed to avoid tight spots in the bagging film.

Before completely sealing the mold, the vacuum port should be added to the bag. There are several different ways to apply the vacuum, but each one falls under one of two methods: either through the bag, or at the bag's seal. To apply a through-bag connector, make a small cut in the bagging film with a utility knife, insert the lower half of the vacuum connector through this hole from the laminate side of the bag, seal the bag, and then fasten the top half of the connector as recommended by the manufacturer.

A vacuum line can also be inserted into the vacuum bag at the seal by wrapping a ring of sealant tape around the vacuum line a few inches from the line's end and then inserting it into the vacuum bag. The sealant tape around the vacuum line can then be tacked to the bag's seal and closed off with additional bits of sealant tape until completely closed. If using this method, be sure to cover the end of the line with bleeder/breather cloth - the bagging film can otherwise fold over the open end of the vacuum line and close it off so air cannot flow out.

Turn on the vacuum pump or connect the vacuum source, and slowly open the valve to the inlet. Check to make sure no bagging film is "bridging" anywhere and that even pressure is evident on the laminate. Ensure that all bagging materials are laying down evenly in the mold and causing only minimal wrinkles before full vacuum is achieved. Throttle open the bleed-off valve as needed to slow down the vacuum as you check for any problems with the bagging system

Again, if the resin is very thin, allow only a partial vacuum to be drawn by using the bleed-

Ensure good resin saturation of all reinforcements…

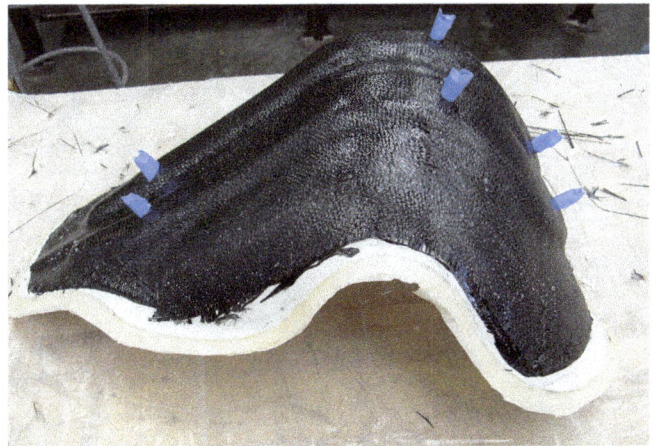

…and then place tape over the exposed fastening hardware to keep them from tearing the bagging film.

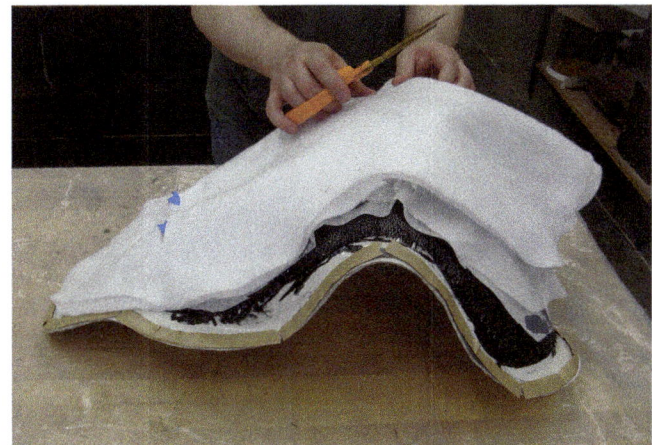

Apply all bagging materials as before…

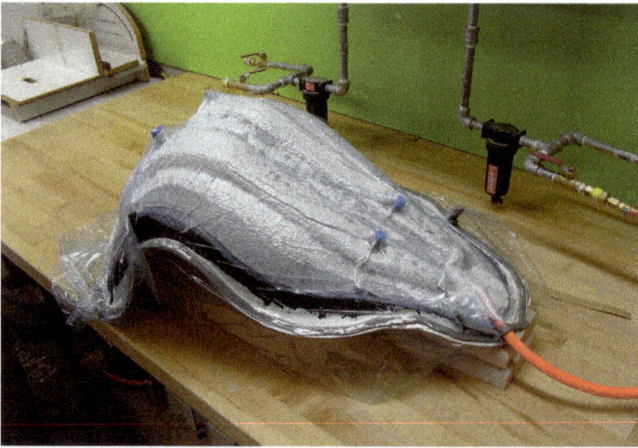

…and evacuate the part. Allow the part to cure.

Use a wedge to remove the cured part…

…and unscrew the bolts from the T-nuts.

off valve in the system. Increase the vacuum again once the resin has begun to thicken, as evident by touching and testing any resin residue that may be on the work table or in a mixing cup left over from the layup.

It is typically a good idea to leave the vacuum pump on long enough to provide assistive pressure to the part until it has cured, even if your vacuum connector is equipped with a built-in shut-off valve. The vacuum source can be removed once the part has cured to a hardened green state (when the resin is no longer tacky), but the part should not be removed from the mold until the curing cycle has completed, as recommended by the resin manufacturer.

After curing is complete, peel off and discard all bagging materials and demold the part. Inspect the final laminate for any imperfections. It is common for the mold-facing surface of vacuum-bagged laminate to have small voids and pits in it, especially when an autoclave (or additional atmospheric pressure) is not used. Satin weave and twill fabrics will help minimize these issues, but a surface coat can also hide small voids, or thicker resin used in the layup will minimize voids as well.

Finally, clean and trim the part, as shown in Handbook – 1. Once completed the lightweight, optimized, and consolidated part will be ready for service.

CHAPTER CONCLUSION

Vacuum bagging processes require several pieces of equipment and specialized materials, but can produce very high quality composites. Practice will help build your vacuum bagging skills, and over time, the process will become second-nature as you extend your capabilities in fabricating superior parts.

Clean up the part and remove any residual PVA from it.

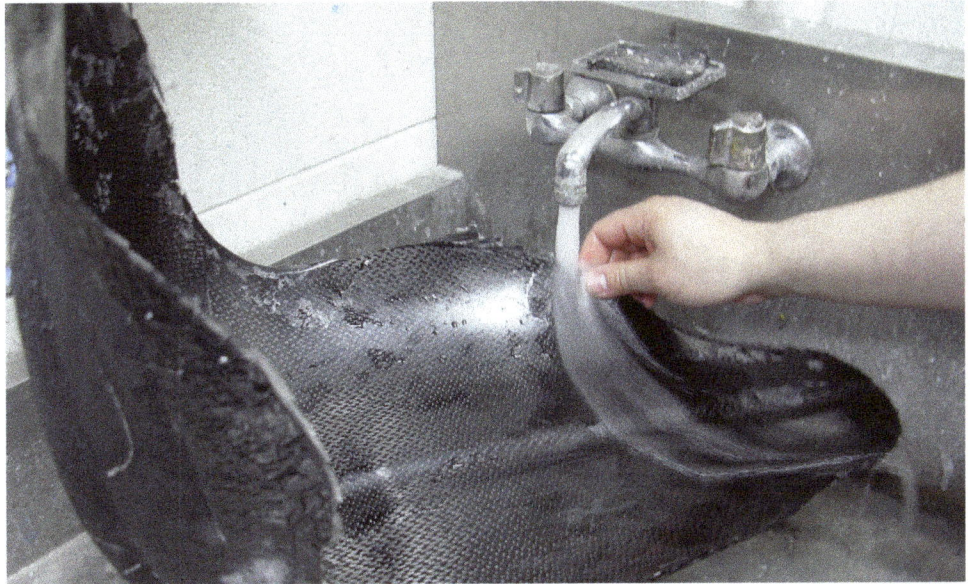

A rotary cutter can be used to quickly trim the rough edges…

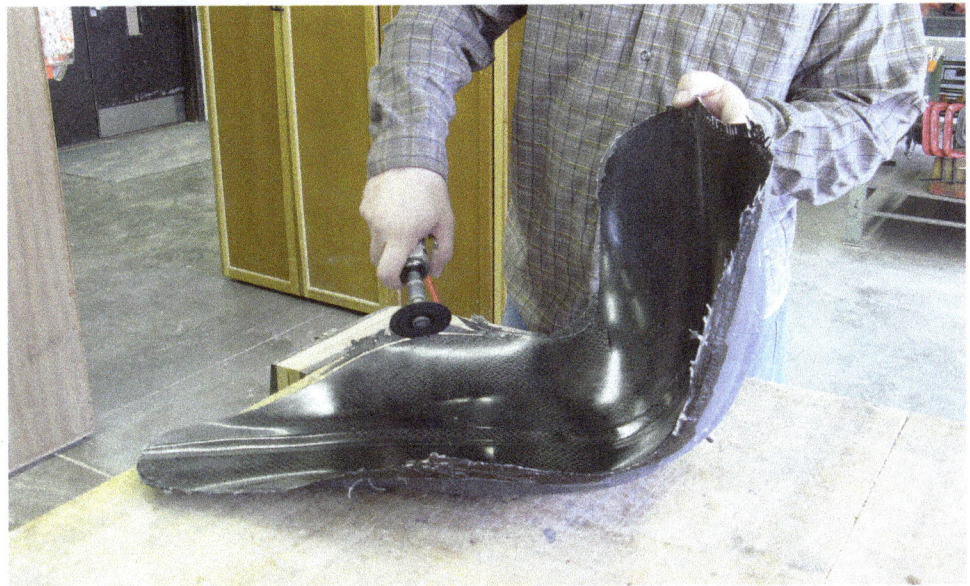

…and the rigid, completed part is now ready for final finishing!

Chapter Five

Pre-preg and Expandable Insert Molding Techniques

Using Pre-preg for Perfect Parts Every Time

CHAPTER INTRODUCTION

Although vacuum bagging techniques work well for parts that are generally open in shape, different molding techniques are required to mold parts with tight or detailed geometry, deep cavities, or that are completely enclosed (as with a "clamshell" mold). Using a mold insert formed from flexible silicone - often called "trapped rubber molding" - can help in this respect. As silicone rubber is heated, it expands, forcing the laminate against the mold under high pressure to produce tightly consolidated, high surface quality composite components.

Low void and high surface quality parts - like the concept car side mirror fabrication demonstrated in this chapter - are possible using heat expandable inserts and pre-preg.

EXPANDABLE INSERT BASICS

While the generic term "silicone" refers to a specific class of chemical compounds, room temperature vulcanized (RTV) silicone is of particular interest in molding composites. This particular type of silicone will cure to a rubbery consistency and a preset durometer (or firmness) when a catalyst is added. It will also accurately reproduce very fine surface detail, is reasonably resistant to the chemicals used in composites, is naturally self-releasing from many resin types, and can maintain its flexibility through a wide temperature range and many layup and cure cycles.

One property of silicone that makes it especially valuable in molding cavity features into composites is its very high coefficient of thermal expansion (CTE), or the rate at which it expands when heated. The thermal expansion of silicone varies with its particular chemical make-up, but will usually expand eight to fifteen times more than aluminum (another material with an inherently high CTE) when heated. Because of this high rate of expansion, heated silicone rubber is especially useful as a means to press a composite laminate against the mold face, creating a heat-activated compression molding system. The silicone can exert up to hundreds of pounds of pressure per square inch - or even more! When used correctly with high-temperature cure composites - such as "pre-preg" (reinforcement fabric pre-impregnated with resin) - the resulting laminate can have excellent consolidation and be virtually void free.

One drawback to using silicone in a mold system is its cost. It is usually sold by weight, and a typical 2-lb container (approximately 28 to 30 liquid ounces) of material can cost between $20 and $40. This usually means that silicone is most cost effective when used only for small parts or, if used for large cavities, contains an inert core or plug within it made of less expensive material (such as a wood, plaster, dense foam, or metal) to fill up space and minimize the amount of silicone needed.

PRE-PREG BASICS

Silicone inserts can be used in a variety of layup procedures, but are especially helpful with pre-preg composites that require high temperatures for curing. Pre-preg composite materials are created in controlled factory settings by pressing resin into dry reinforcement fabric between large metal rollers. Because of the processing accuracy of this impregnation method, pre-preg comes from the manufacturer with the "perfect" amount of resin in it - typically between 60-65% fiber and 40-35% resin. Pre-preg is used almost religiously in aerospace and military composites

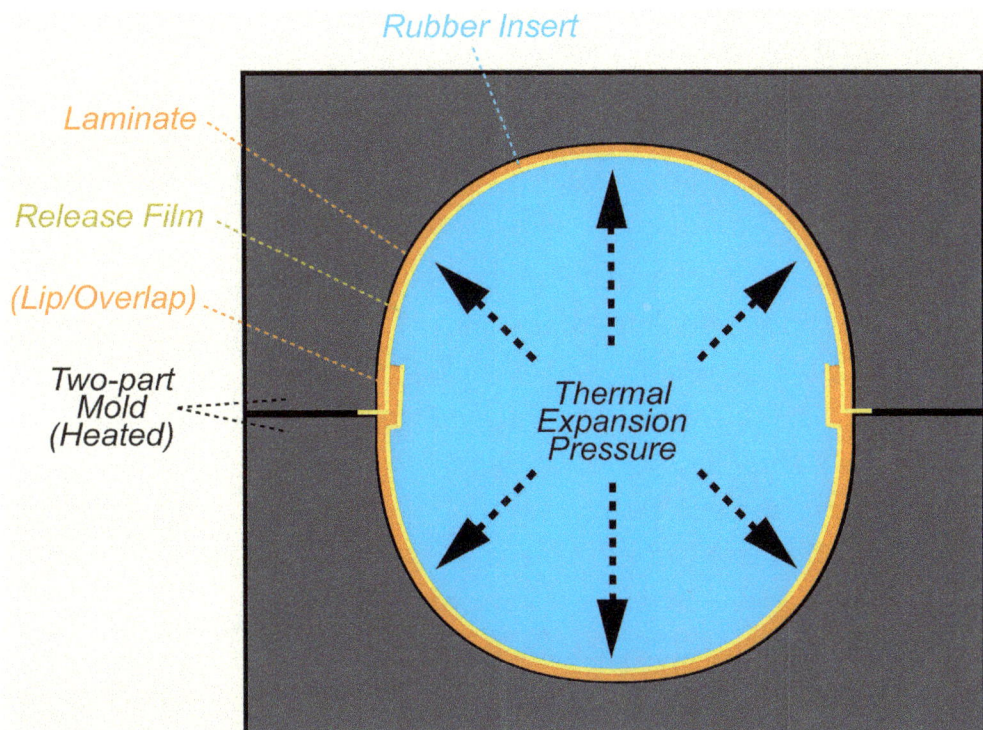

"Trapped rubber molding" uses heat to expand a molded rubber insert, which compresses the laminate against the mold walls.

Pre-preg and Insert Lamination Schedule

For Dummy Part:
Resin System: Epoxy
8 Layers - 8 oz. Fiberglass (Plain Weave)

For Final Part:
Resin System: Pre-impregnated Epoxy
7 Layers - Carbon Pre-preg (3K - 2x2 Twill)
Additional Selected Reinforcement Layers

Materials and Supplies

7075 Aluminum Billets
3-Axis CNC Milling Equipment
Silicone RTV (10 - Shore A Hardness)
Bagging Film and Release Ply
Sealant Tape
Scrap 5.7 oz. Carbon Fabric (3K - 2x2 Twill)
Mold Release/Parting Wax
Modeling Clay
Mixing Cups, Sticks, and Paintbrushes
Rubber Spreaders and Heat Gun
Disposable Gloves, Clean Rags/Cloths
Ruler, Silver Marking Pen, Tracing Paper
Scissors and Miscellaneous Hand Tools
Masking Tape and Sandpaper
Rotary Tool with Cut-off Wheels
Miscellaneous Fasteners/Hardware
Vacuum Source and Vacuum Tubing
Oven and Digital Thermometer

The pre-preg and expandable insert lamination schedule, along with materials and supplies, used in this demonstration.

because of their optimum fiber-to-resin ratio and ability to be laid in a mold with very little mess and maximum control over fiber orientation. These pre-preg materials are supplied in rolls (similar to raw reinforcement fabrics, yet with plastic backing on each side), and have a slightly tacky surface reminiscent of a fruit roll-up. They generally require special curing cycles in an oven or autoclave to fully harden, but will yield a laminate with excellent properties.

When it comes to fabricating parts, pre-preg lays up differently from the raw fabrics and resins used in a wet layup. Pre-preg is significantly less messy and easier to handle and laminate. Fabricators who are used to wet layups are usually astounded at how easy (and clean) pre-pregs are to use. Pre-preg is, however, slightly more difficult to drape in a mold, requiring warming with a heat gun and finessing with a rubber squeegee and one's fingers to control the fabric weave so it will lie onto complex surfaces. Warmed pre-preg will also get tackier than wet fabrics and can grab onto rubber gloves during handling - a property of pre-pregs that can be particularly annoying since gloves must be worn in order to keep skin oils from affecting the bond between pre-preg plies.

Because of pre-preg's tackiness, it is not advisable to use PVA on the mold surface since pre-preg will grab and peel up the PVA if the ply is repositioned. In place of PVA, other specialty releases (such as water-based types) are recommended. While wet layup plies are usually gooey enough to slide over or lie onto a mold face, pre-preg must be pressed into place to work out air pockets and bubbles and forced to conform to a mold. This property of pre-preg materials actually makes them easier to place correctly in a mold, allowing for repositioning as needed until the precut pieces fit correctly. Once positioned as needed, the pre-preg laminate is then pressurized and heated. The resin matrix becomes slightly more liquid with heat, allowing resin to flow between the laminate plies, and remaining air bubbles or voids to be pressed or drawn out of the laminate.

Pre-pregs can be trimmed much more cleanly and precisely than wet layup plies. It cuts very easily with a utility knife or scissors and doesn't produce frayed edges or leave stray, furry fibers that can fall into the layup. In fact, trimming uncured pre-preg is very similar to "green trimming" a wet layup because its resin matrix is actually in a "B-stage" (or "green" state) of cure. This property of the pre-preg makes it possible to easily pre-cut (or "kit") patterns prior to layup - something that can greatly speed up a layup procedure. In industry, entire laminate schedules are planned and cut into patterns, assigned part numbers, and systematically arranged for layup and speedy production.

Until recently, pre-pregs were out of the reach of most small fabrication shops because of both cost and availability issues. Compared to raw fabrics of the same weight and weave style, pre-preg fabrics can cost nearly three times as much (even with the cost of resin factored in), they have a limited shelf life (after which they will cure and harden on their own), and are

Clamshell molds for this demonstration were created from 7075 aluminum billets reclaimed from a local metals salvage yard (www.alreco.biz).

3-axis CNC milling machines can produce very complex, high accuracy molds. This is the top half of the clamshell mold set being milled.

Mold surfaces were hand sanded and then wet-sanded to 400-grit to remove machining scallops and marks.

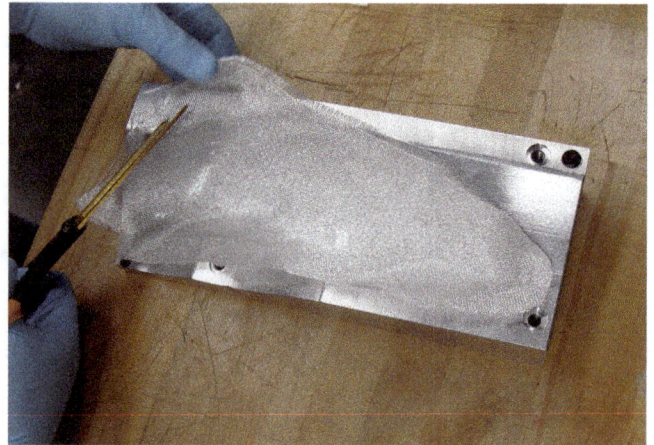

These surfaces were then hand polished to a high shine.

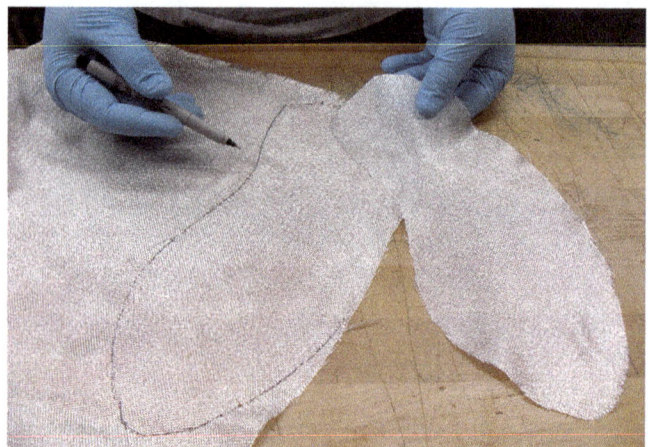

The finished molds (with alignment and sealing hardware), ready for layup. Total machining and finishing time for these molds was roughly 60 hours.

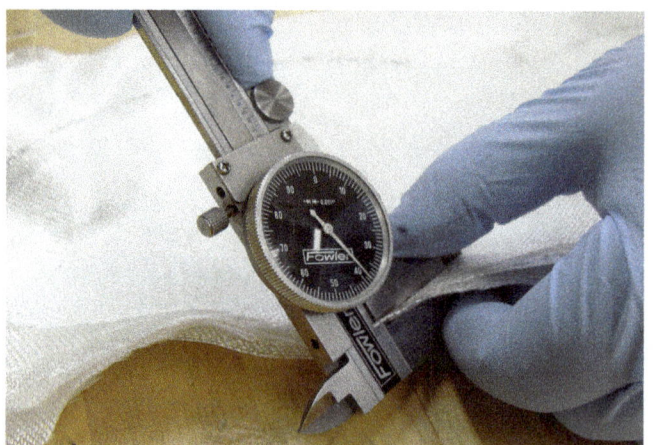

To make the "dummy" part, cut fiberglass fabric to fit the shape and contours of the mold.

Use this cut fabric to pattern additional plies.

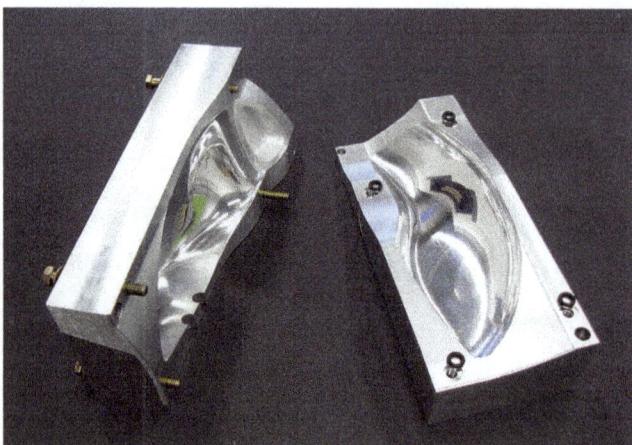

Measure these plies' thickness with calipers, adding more plies until the desired part thickness is reached.

Cut all the plies in preparation for layup. These will be vacuum bagged in the mold for good consolidation.

Layup the dummy part in the mold using the cut fiberglass plies.

Wax the mold with paste wax and PVA.

Trim off excess fabric…

Apply masking tape around the mold's edge to aid in post-layup cleaning of the sealing flange.

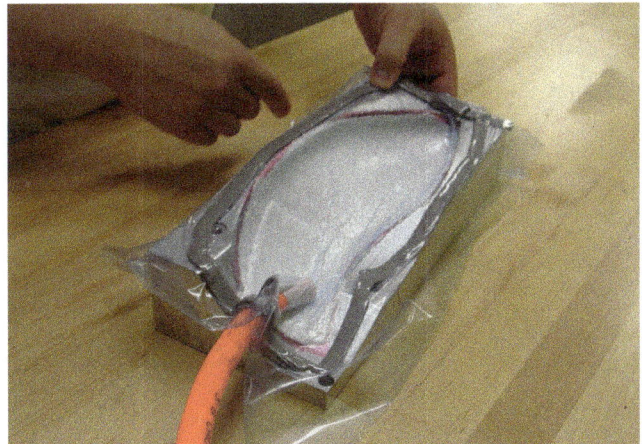

…and bag the part (as shown in chapter 4). Apply the vacuum, check for leaks and allow the part to cure.

After cure, remove all bagging material and sand down any wrinkles or ridges in the dummy part.

Apply modeling clay to further smooth the inner surfaces, if needed.

The finished dummy parts are now ready to form the silicone insert.

available from very limited sources (which are sometimes hard to locate). Even though more composites suppliers are now making pre-preg available to the do-it-yourselfer, pre-preg's use continues to be limited by cost and unfamiliarity. In spite of these hurdles, though, the full potential of pre-pregs can be unleashed with a little practice.

EXPANDABLE INSERT AND PRE-PREG FABRICATION TECHNIQUES

To demonstrate the use of heat-expanded silicone inserts and pre-preg in a molding procedure, we will create a driver's side mirror for a concept sports car. A "clamshell" type aluminum mold and high-temperature curing pre-preg composed of carbon fiber and epoxy will be used and explained throughout this chapter. Given the precision needed to ensure good mold alignment, and the high temperatures required to cure the pre-preg (between 270 and 310 degrees Fahrenheit for the pre-preg used in this demonstration), the mold was cut from billets of 7075 aluminum with a 3-axis CNC mill after being designed using 3D computer modeling software. This method of computer-modeled and CNC milled mold making requires a bit of technical skill, but is required for the good part accuracy and complex curvature needed with this particular part. Even so, several CNC-capable shops offer computer modeling services that make it possible for the do-it-yourselfer to still employ these mold making methods.

After machining was complete, the resulting two-part mold was sanded smooth and then polished by hand to a mirror finish. Alignment and clamping hardware were next added to lock the two mold halves shut during oven curing of the laminate.

This chapter also demonstrates a method for creating the silicone insert itself, along with a simple lip-forming method to create two fitted halves to the composite side mirror. This particular two-piece design will make it possible to mount a power mirror mechanism within the side mirror body, which can then be secondarily bonded with epoxy paste and sealed shut.

Add o-rings to seal off the threads of the alignment/clamping bolts in the mold.

Tighten the bolts so they are snug, but not overly tight.

Use clear packing tape to seal off the edge of the mold so the silicone will not seep out of the mold during casting.

Mix up the silicone resin, as directed by the manufacturer. An accurate scale is imperative to a proper mix.

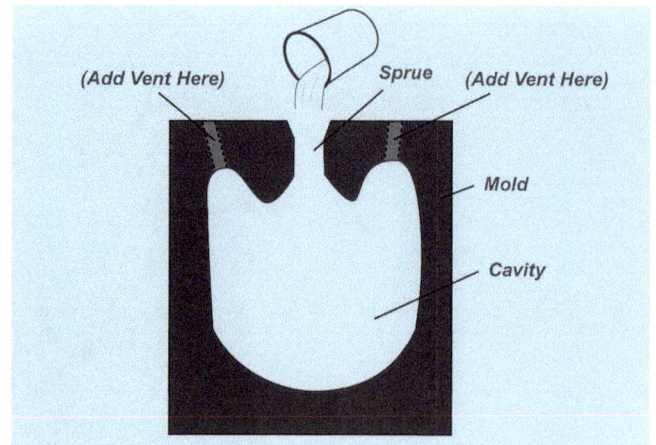

Odd shapes may need additional "vent holes" so air will not become trapped during the silicone pour and cause large voids in the casting.

Mix the silicone thoroughly, scraping the sides as needed…

After cure, open the mold, and remove the dummy part and silicone insert.

…and pour the silicone into the mold sprue. Note that this mold has been tipped so the silicone can be poured into the highest point of the mold cavity.

Remove any silicone flash that may have worked itself between the mold halves.

Alternative methods could be also used to create such a small hollow part (i.e. such as with internal bladders, or vacuum-bagging and then secondary bonding), but for low production runs with high-temperature pre-preg, silicone inserts can be much less time intensive overall and yield excellent results.

Based on tests with some carbon and epoxy composite test samples, a wall thickness of .060" (about 1/16") will be used for the side mirror design, with some extra reinforcement added in key locations. The mold system must accommodate this part thickness by leaving a gap of this same size between the mold and the silicone. This can be done by laying up a "dummy" part in the mold of the same thickness as the final pre-preg part and then filling the new inner cavity with silicone. Inexpensive fiberglass can be used to form the dummy part, but vacuum-bagging is recommended to ensure uniform thickness of the part over the mold walls - a poorly sized dummy part will form a poorly sized silicone insert, and thus a poorly or unevenly consolidated pre-preg part. A thin layer of plasticene clay, multiple layers of painters tape, or special adhesive-backed wax sheet (available from some composites suppliers, such as www.freemansupply.com) could also be used instead of

Remove the dummy parts from the silicone insert. Silicone is self-releasing, so the parts should come off without too much effort.

The completed dummy and silicone parts. Store the dummy parts in a safe place in case you need to form a new silicone insert in the future.

Close off the sprues to prevent over expansion of the silicone insert later during heating. This 18 gauge sheet metal piece is held in place by the pre-preg during layup.

..and trace it onto tracing paper.

To create a properly sized pattern on complex surfaces, lay in and mark the edges of scrap fabric with a similar weave and weight to that found in the prepreg.

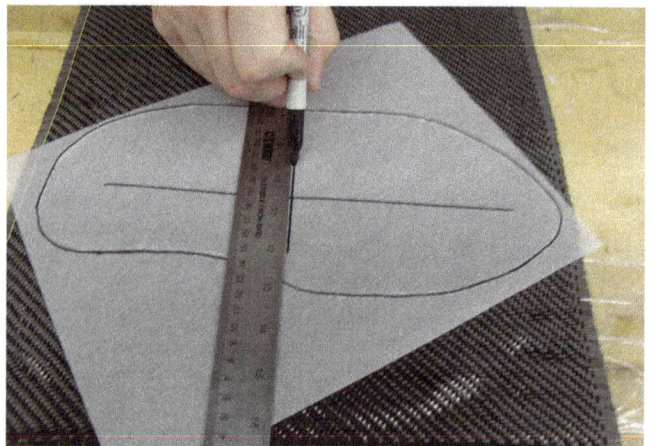

Be sure to mark the orientation of the fibers in the fabric so the pattern can be aligned with the corresponding fibers in the pre-preg material.

Remove this piece, straighten out the weave...

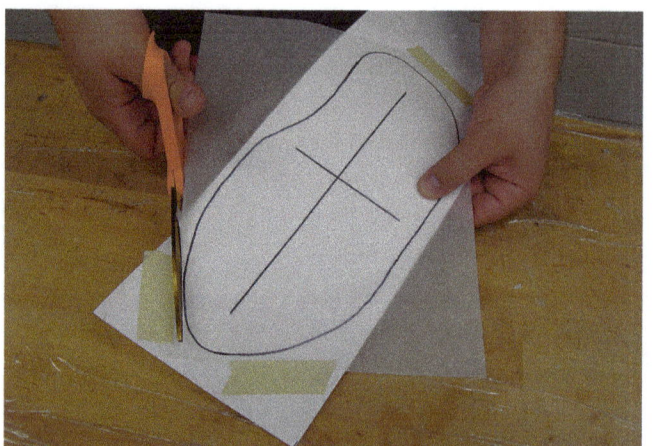

Transfer the traced pattern to poster board.

The cutting pattern is now ready.

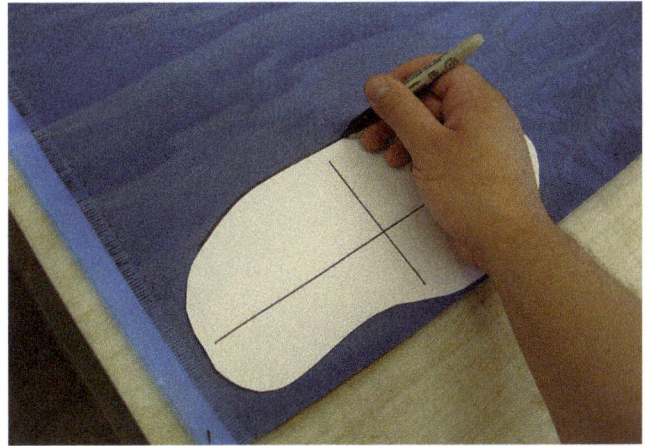

Trace the cutting pattern on the pre-preg and cut wisely to avoid waste—this material is expensive!

If using refrigerated pre-preg, leave the pre-preg on the roll in the bag until it has warmed up to room temperature—this will prevent condensation from forming on the pre-preg.

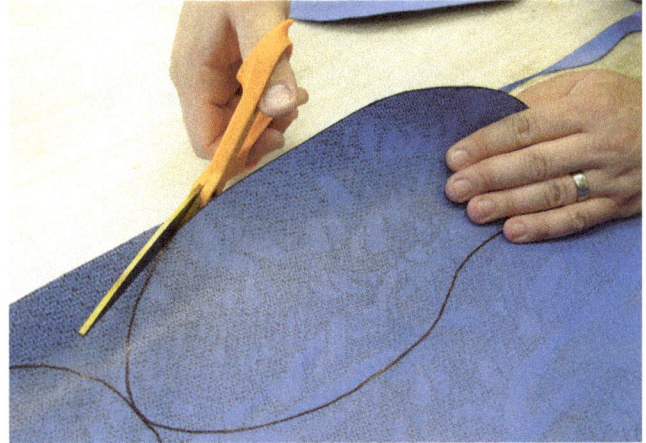

Cut the pre-preg using either scissors or a sharp utility knife.

Pre-preg has plastic backing on both sides of the material.

Use acetone to clean off all surfaces of the mold. Apply a water-based release agent over ALL surfaces of the mold. Such releases can better handle the high temperatures used in pre-preg processing.

Lay out all the plies and supplies needed for the pre-preg layup.

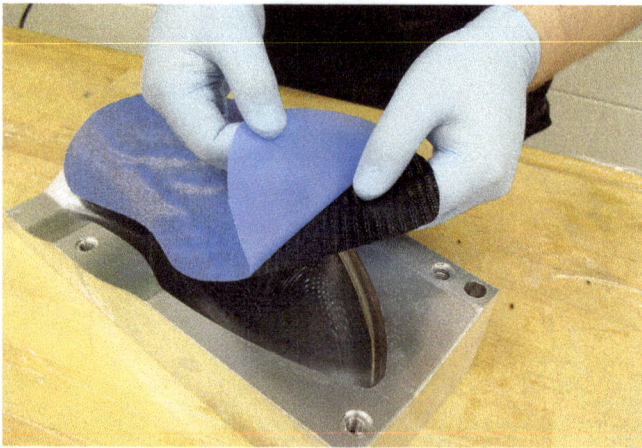

With gloved hands, peel off the backing from the pre-preg.

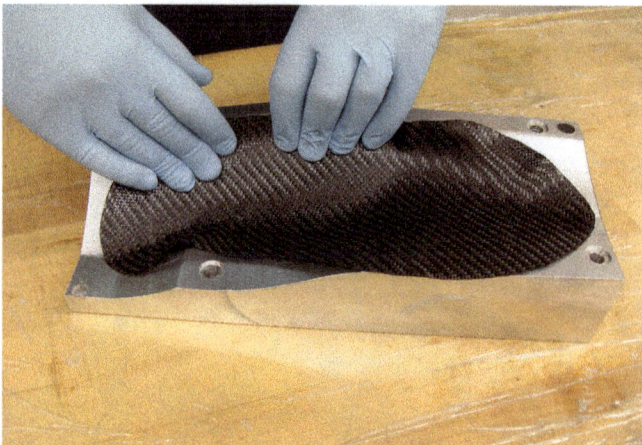

Press the pre-preg onto the surface of the mold firmly and completely.

fiberglass to create the thickness of the dummy part.

After the dummy part has cured, trim off any flash from it and use body filler, spot putty, or clay to fill any voids until they match the thickness of the part as closely as possible. Close and bolt the mold halves shut, and then seal the seam with clear packing tape to keep the silicone RTV from leaking out. Leave a hole (or "sprue") on top of the mold through which the silicone RTV can be poured. It is normal for small air bubbles to remain in the RTV after mixing. In fact, a void free RTV casting is only possible with a large vacuum chamber made specifically for degassing mixed fluids. However, most large air bubbles can be avoided by ensuring that there are no areas in the mold that may create pockets, such as those that may be higher than the sprue opening in the mold.

The silicone resin used in this demonstration is a tin-cure type with a 10-Shore A hardness and a service temperature up to 340 degrees Fahrenheit, (available from Plasticareinc.com). Mix up the silicone RTV as directed by the manufacturer, making sure to follow the recommended mix ratios exactly - just as with many other thermoset resins, silicone RTV may never completely cure if the mix ratios are not correct. Mix the silicone completely, scraping the sides and bottom of the mixing cup while stirring so the resin will be catalyzed thoroughly. Once mixed, pour the resin in a thin stream into the mold until the cavity within the dummy part has filled completely. Allow the silicone RTV to fully cure for as long as recommended by the manufacturer.

After the silicone has cured, remove the tape from the mold seam, open the mold, and carefully remove the newly formed silicone insert and dummy part from the mold. Trim off any flash from the silicone that may have worked its way in between the mold halves. If the geometry of the part requires it, the silicone insert can be cut into sections for easier removal using a sharp utility knife. The silicone insert is now ready to expand as the mold system is heated and compress the pre-preg.

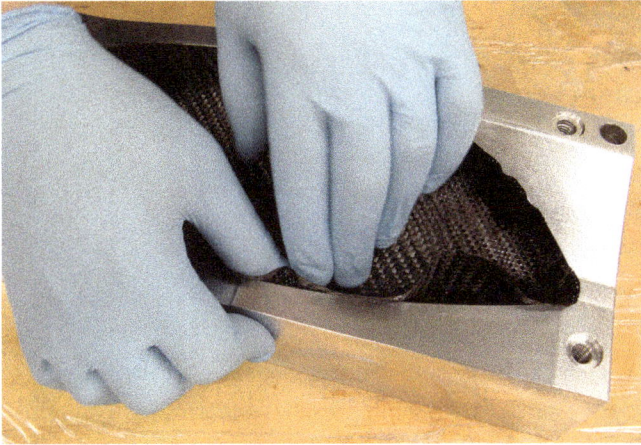

Carefully push the fabric into tight contours.

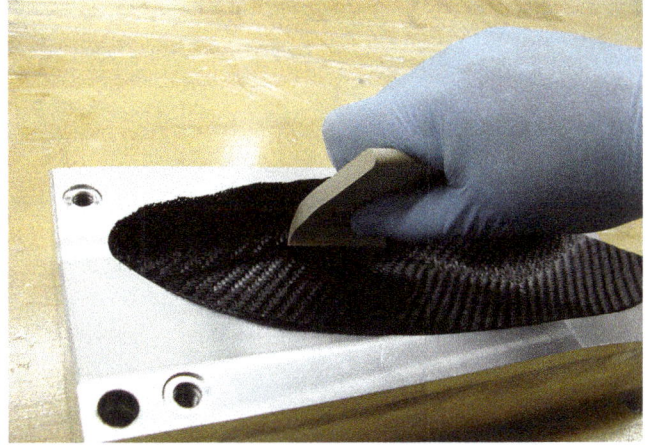

As layers build up, a rubber scraper can be especially helpful in pushing the pre-preg into tight corners.

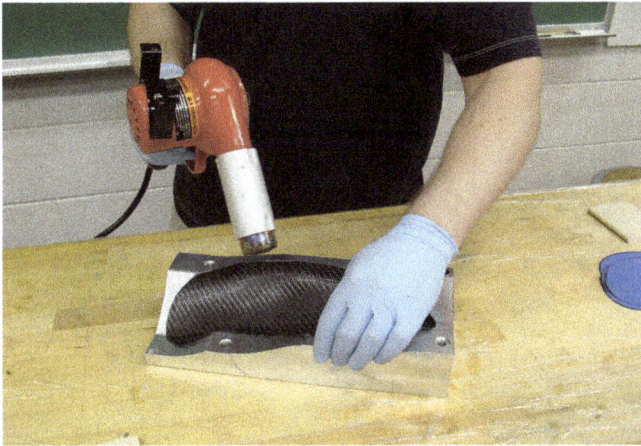

A heat gun can further soften the pre-preg so it will conform more easily to complex curves.

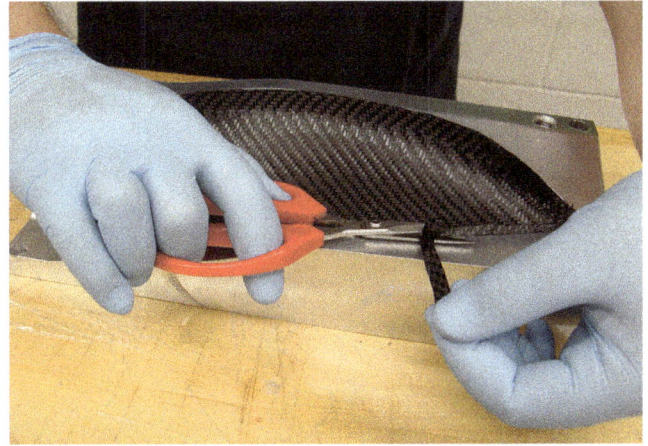

Trim off any excess edges using sharp scissors (small scissors tend to work best on parts of this size).

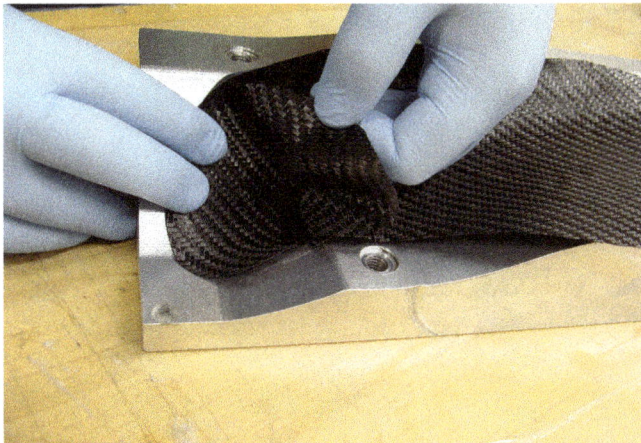

Add successive plies to the mold, as listed in the lamination schedule.

The two mold halves ready for silicone insertion. Note the flaps left on the lower mold half for later secondary bonding between the two parts.

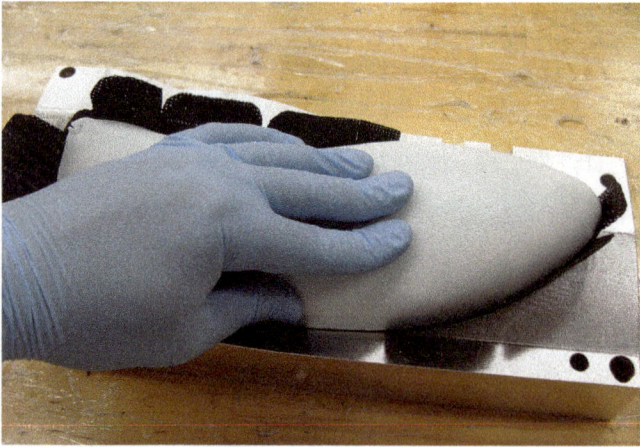

Press the silicone insert into the mold.

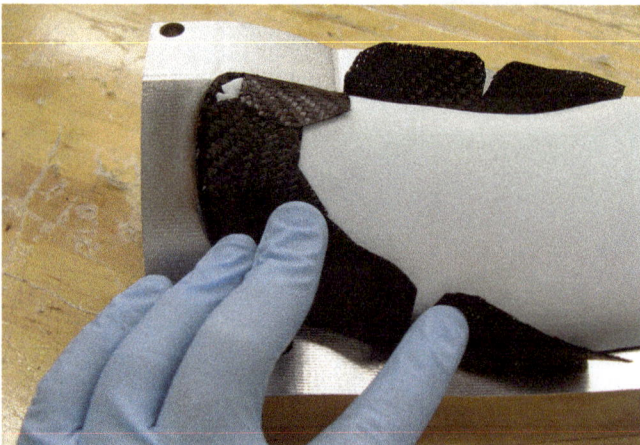

Fold over any flaps that will be used for secondary bonding.

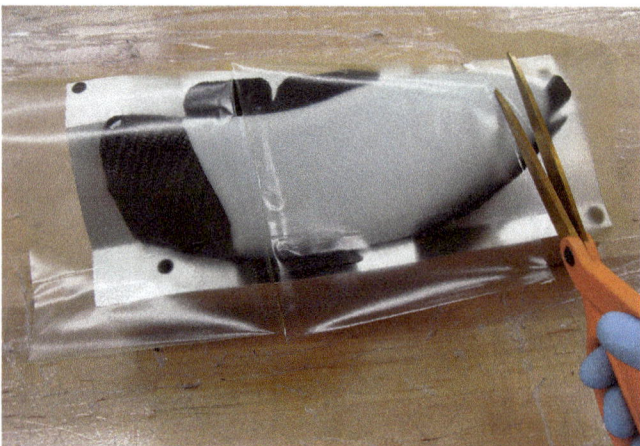

Cover the insert and flaps with high-temperature nylon film...

Prior to layup, the holes and sprues in the mold must be closed. Failure to cap these holes will cause the silicone to actually expand out of the mold and relieve pressure from the laminate. Even worse, this could completely ruin the silicone insert, possibly requiring a complete re-fabrication of this costly mold component.

Pre-preg can now be laid up into the mold after it has been cut and prepared. The pre-preg used for this demonstration was an epoxy matrix, room temp storage, and high temp cure type with a carbon 3K 2x2 twill weave fabric (available from www.fibreglast.com). As previously mentioned, kitting can speed up the layup process. The kitted plies can be created by first cutting patterns that follow the shape of each pre-preg piece to be laid in the mold. These patterns can also be labeled to show each ply's order of layup (as determined in the lamination schedule), and the intended orientation of the ply's fibers. For relatively flat composite parts, patterns can be made by laying paper or poster board onto the mold surface, and then trimming them to the size of the ply, as needed. However, this technique of patternmaking works only marginally well on complexly curved surfaces. Patterns for more complicated molds can be made by using a fabric of a similar thread weight and weave to that found in the pre-preg. This raw fabric can then be placed in the mold, formed to match the curvature of the mold, marked, removed and flattened, and then traced to create the correct ply shape. Although pre-preg can be heated and formed to fit highly curved and detailed molds, the fabric weave should not be opened so much that the part will be weakened. When a part is too curved for practical fabrication from a single continuous piece of pre-preg, multiple pieces can be laminated together, as long as the edges of the plies overlap well and are reinforced by successive layers.

Laminate the pre-preg plies one at a time, pressing them onto the mold face with firm pressure. The first layer of pre-preg is always the most difficult to place because it will tend to pull away from the release-coated mold as it is pressed and formed to the mold's curvature. Be sure to work out all bumps, bubbles, wrinkles and lifted sections in the pre-preg, as these will translate to successive layers and cause voids and imperfections in the lamination. Place each additional ply of pre-preg in the mold with care, repositioning it as needed before pressing it to the previous layer. Laminating these plies of pre-preg will feel somewhat like placing pieces of double-sided tape on top of each other - they will not easily slide into position, and they will take a bit of work to separate if firmly pressed together.

Once the pre-preg has been laminated in the mold, place the silicone insert into the mold, and seal off the mold halves from each other with two layers of nylon bagging film. This film will be able to tolerate the curing heat of the pre-preg, but will keep the two composite pieces from fusing together and permanently trapping the silicone insert within them. The mold can then be closed and bolted shut, making sure that grommets are used to seal off the bolt threads from any resin that may be pressed out of the composite during heating and curing.

With mold systems like this, heat can be both a friend and an enemy. Heat will cause expansion of most materials, but this can be very problematic when a mold and the parts formed in it are of differing materials. When heated, a mold will expand, and the pre-preg laid in it will conform to the larger dimensions of the heated mold. After curing, the mold will cool and contract, causing buckling or residual stresses in the part if this mismatch of thermal expansion is too great. Since this particular mold system uses a highly expanding aluminum mold, the lowest possible heat and maximum cure time is recommended (about 270 degrees Fahrenheit for four to six hours) to minimize contraction problems between the mold and the carbon fiber.

Adjust cure times according to the thickness and thermal conductivity of the mold and the laminate in it. Thick, insulative molds will take significantly more time to warm and cool than thinner ones. For example, a sheet metal mold will be able to heat up and cool down very quick-

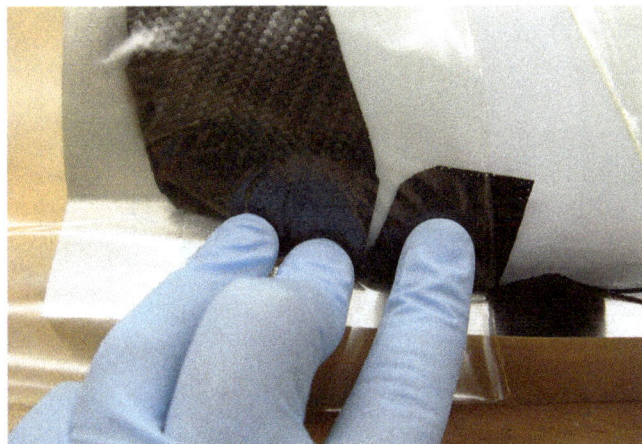

…and press the film onto the pre-preg flaps to hold it in place. Put an additional ply of nylon film over this one to aid in demolding.

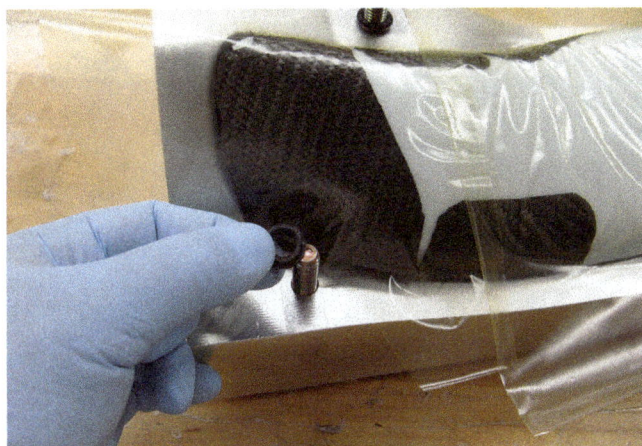

Again, place o-rings over the threads of the alignment/clamping bolts.

Close the mold and snugly secure the bolts.

Place the mold set and laminate into a shop oven...

...and open the mold halves...

...and follow the ramp-up and ramp-down recommendations for the material. A digital thermometer can accurately monitor the "real" temperature in the oven.

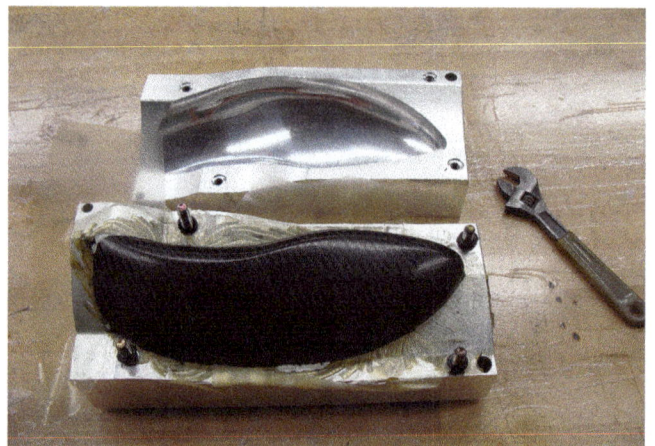

...to reveal the molded, cured parts.

After curing, remove the mold bolts...

Peel off the nylon film...

ly, whereas a thick, insulative plaster mold may take significantly longer. In production, fabricators commonly use thermocouples strategically placed over the laminate or built into the mold to accurately monitor the temperature of the mold/laminate system. To account for the mass of the large aluminum mold used in this demonstration (which is very thick, yet highly conductive), the cure time was extended by about 50% beyond that recommended by the pre-preg manufacturer. Use a clock and thermocouple/thermometer to follow the recommended temperature ramp-up (about 5 degrees Fahrenheit per minute for this particular pre-preg) and ramp-down (again, about 5 degrees Fahrenheit per minute) as strictly as possible. Since it is impossible to "over cook" a pre-preg, err to the side of too much time at the proper temperature, if possible.

After allowing the mold and part to heat, cure, and cool, it can be quickly demolded and the part can be put into immediate use. As the pre-preg plies were carefully trimmed and formed during layup, they may need very little (if any) final trimming once cured and demolded.

…and remove the silicone insert.

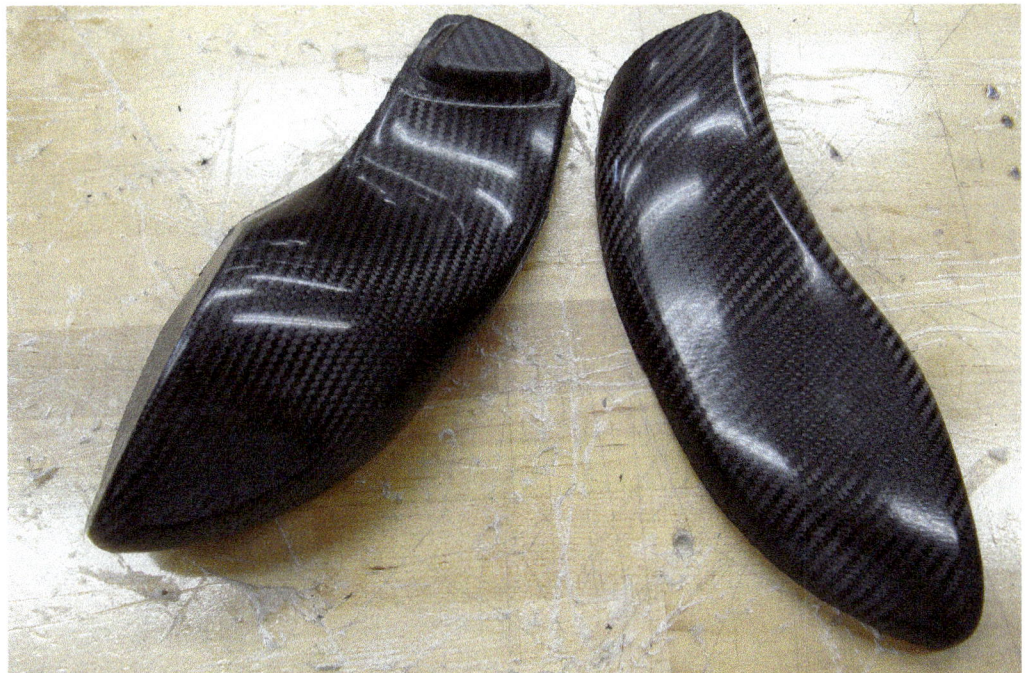

The final parts will have an excellent surface finish and tight consolidation.

A close-up of one part shows how nicely the laminate turned out.

CHAPTER CONCLUSION

Once a suitable mold has been fabricated, lamination of pre-preg plies is relatively straightforward and easy to perform. The combination of desirable properties between the pre-preg (low void, excellent resin-to-fiber ratios, etc.) and the silicone insert (heat expandability, flexibility, natural release, etc.) will be evident in the quality of the resulting part - one that is lightweight, strong, with superior surface quality, and tight laminate consolidation.

The two halves can now be closed together using the perfectly molded bonding flaps.

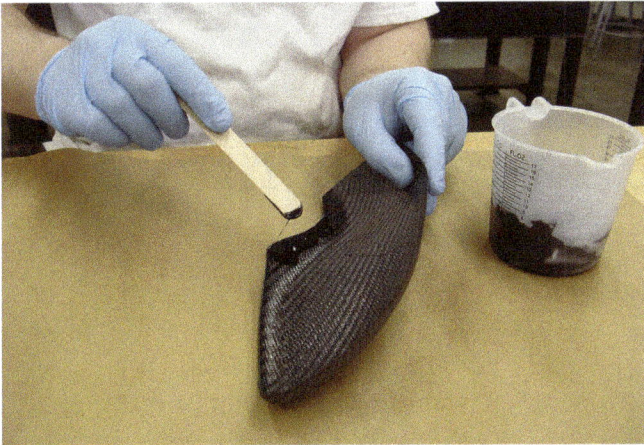

Apply thickened, pigmented resin to the bonding flaps of the parts…

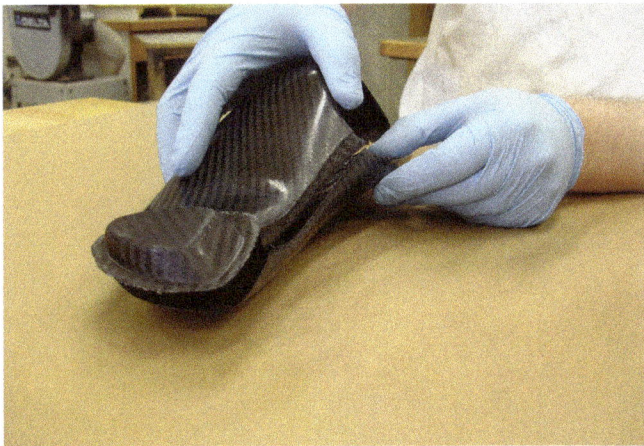

…and join them together. Further filling and sanding will produce a very clean seam.

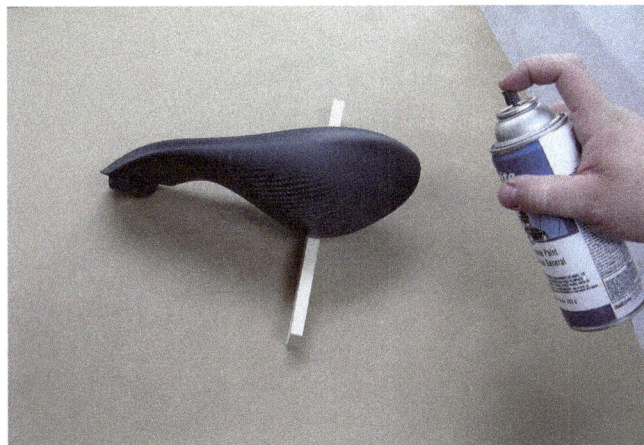

The part can then be sprayed with a clear lacquer to enhance its shine. Secondary bonding of the mirror mounting mechanism inside will complete the part.

To "hide" the seam, black paint can be sprayed on it and feathered out from the seam.

Chapter Six

Inflatable Bladder Molding Techniques

Creating Seamless, Hollow Composites

INTRODUCTION

Another method of applying pressure to mold a laminate is to use an inflatable "bladder" - or an air-filled membrane that forces the laminate against the mold surface. An inflatable bladder system is typically used to mold shapes that are hollow but that cannot accommodate a removable rubber insert or vacuum-bagging arrangement.

More generally, though, a bladder could be used for any closed mold system that needs high pressure without an autoclave, or that may even require more pressure than an autoclave can safely provide. This chapter will describe some of the techniques used to create such bladder-formed parts.

Enclosed, seamless composite parts can be made by inflating a "bladder" within them - as this chapter demonstrates with a carbon fiber gear shift knob.

INFLATABLE BLADDER BASICS

Inflatable bladders are especially helpful in molding seamless or complex parts with high surface detail or smoothness. However, unlike most other molding methods, a bladder may be designed to remain in the part after molding, or to be pulled out and completely removed. Aside from the use of the inflated bladder, layup and demolding are the same for bladder-formed parts as for most other molding techniques.

One caveat to using inflatable bladders is that if they are poorly designed, allowed to expand haphazardly, or are inserted incorrectly, they may over-stretch and burst. A bladder can be constructed from any of a variety of materials, but it must be flexible, tough, airtight, and compatible with the resin in the composite. Likewise, it must be able to withstand high pressures without rupturing and designed so that it will conform to the mold shape as uniformly as possible when inflated. The best way to avoid such failure of the bladder is to keep the bladder from stretching too much by making sure it evenly conforms to the shape of the mold from the very beginning. The primary purpose of the bladder should not be to stretch and press the composite against the mold (though some stretch can be helpful), but rather to create a barrier that seals off the pressurized air from the laminate as it compresses the composite against the mold. If pressure forces the bladder into a deep cavity or unusual shape in the mold that over stresses the bladder, it may simply pop and lose all usefulness. Some materials that fit the requirements for a bladder include nylon and polyethylene films, latex, and silicone.

Nylon and polyethylene films are inexpensive, come on rolls, and can be fabricated into a wide variety of bladder shapes. Nylon film, though less flexible, is generally best used for high temperature applications. Once cut into shape, typical vacuum bag sealant tape can be used to close and seal the edges. When using these films to make bladders, it is best to slightly oversize them since they will not stretch nearly as much as some other bladder materials.

Latex rubber is naturally flexible, stretchable, and inexpensive, making it an excellent choice as a bladder material. It is available in sheets, tubing, and premade forms, by some specialty manufacturers who offer custom latex bladders designed expressly for composites applications. Custom latex bladders can be costly for low production or one-of-a-kind parts, but off-the-shelf latex forms (such as thin-walled tubing, party balloons, or even plain condoms) are still usable if the mold system is designed specifically for them. Latex bladders are notorious for inflating irregularly and forming bubbles in sections that are composed of

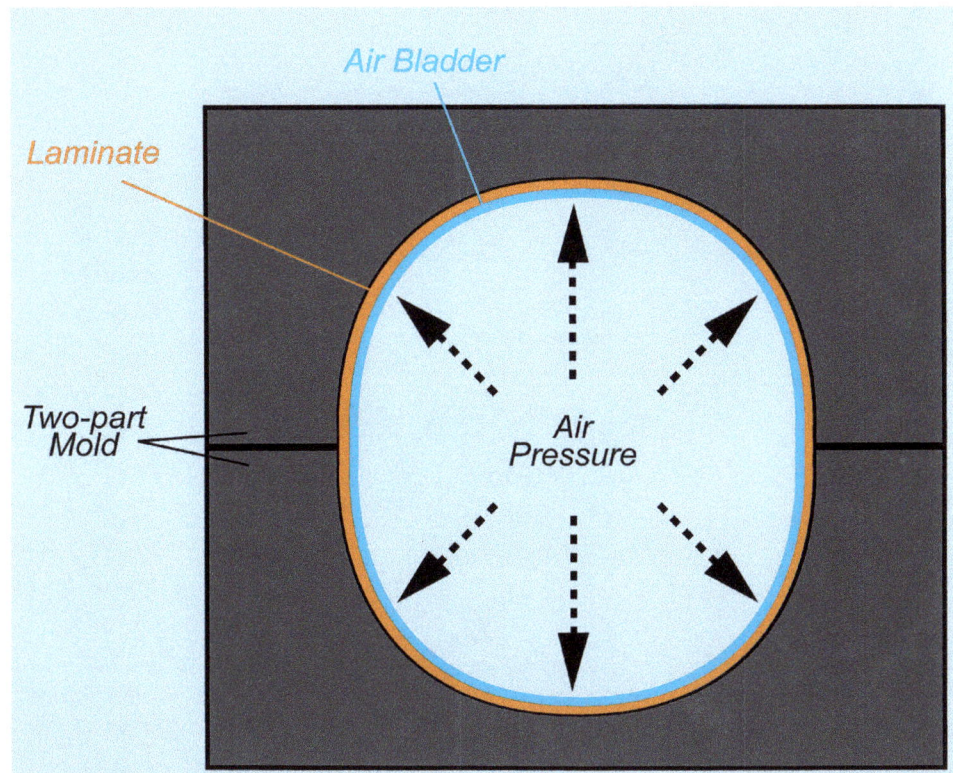

Inflatable bladders are used to form hollow, seamless parts by using pressurized air or fluid.

slightly weaker material. A long clown balloon is a good illustration of this, since it will tend to inflate in one section first, and then fill up the rest of the way rather than filling uniformly all at once. Such bubbles will also tend to grab sections of a laminate as it inflates, keeping the bladder from expanding freely and uniformly, and thus allowing over-inflation in other areas of the bladder or uneven tugging on the laminate as the bladder expands.

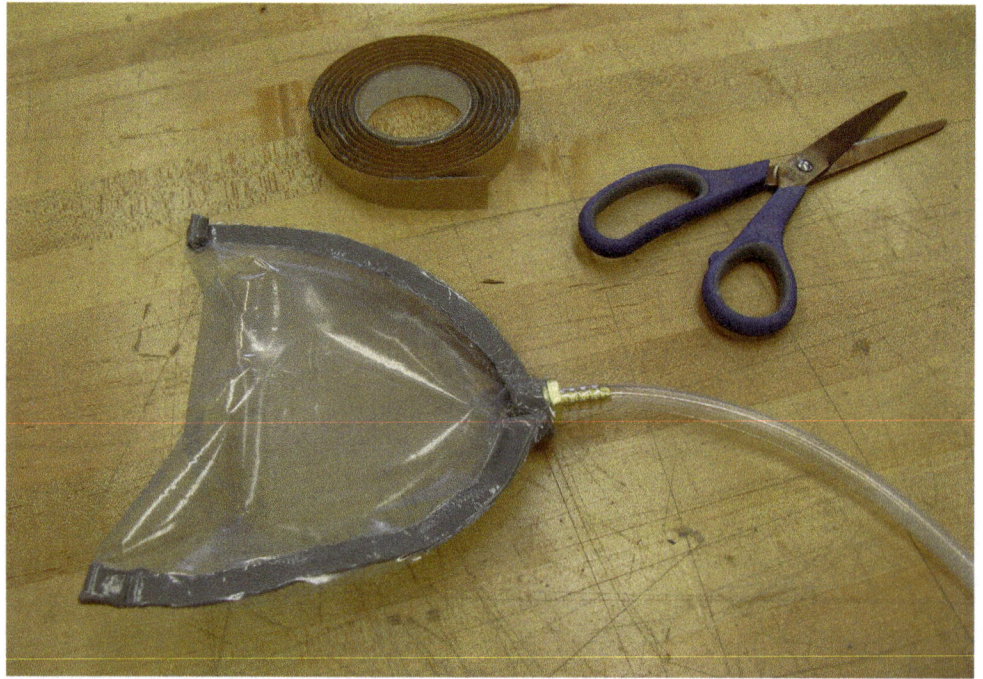

Custom bladders can be made from flexible film materials and sealant tape.

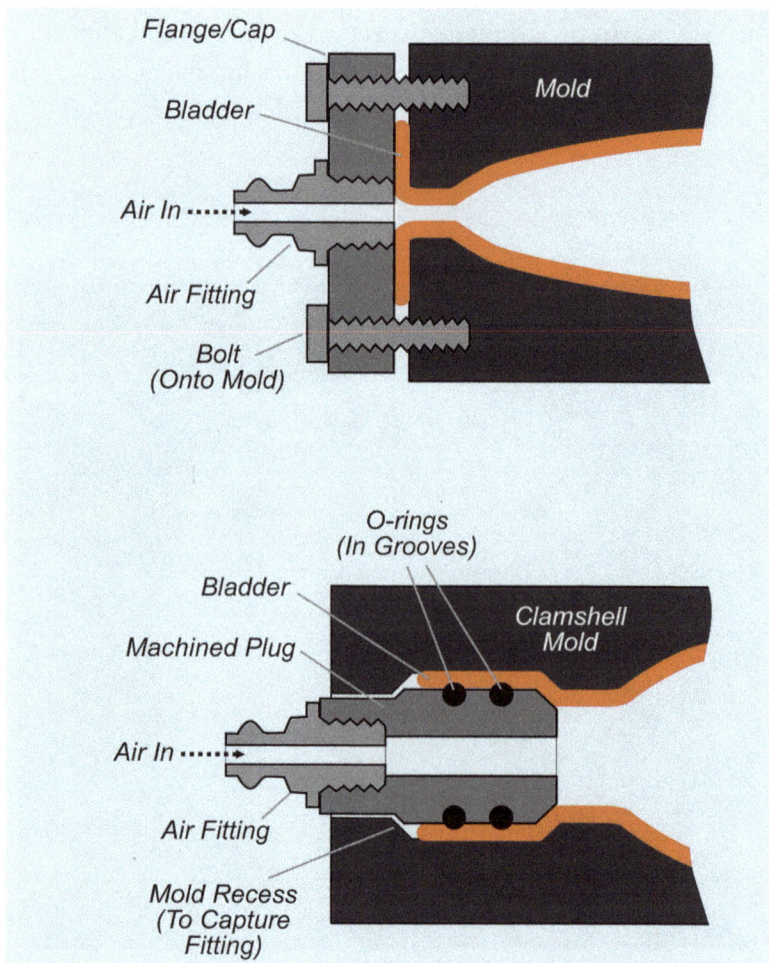

Some examples of sealing flanges used with inflatable bladders.

Silicone is another flexible material that can be used to form an inflatable bladder. Bladders of this type can be brushed onto a mold surface, cured, removed, and then used multiple times. Silicone's natural release makes removal from the laminate easy as well. For more rigorous applications, the silicone may even be strengthened with reinforcement cloth to minimize tearing of the silicone through repeated or rough use.

When designing the actual inflatable bladder system, there are a few things to consider. First, the bladder should match the shape of the mold as closely as possible to avoid over-stretching and rupturing the bladder. Second, air delivered into the bladder should be tightly sealed and the pressure should be controlled. Lastly, the strength of the mold should adequately handle the pressures exerted on it by the bladder.

Flexible bladder materials have varying amounts of elongation (or stretch), with latex and silicone having some of the highest elongation available. Yet any flexible material will have a point beyond which it will tear or rupture. It is important to remember that, a bladder that is already very close in shape to the mold will be in no danger of over stretching and will be considerably more reliable than one that is not.

One troublesome aspect about using a bladder is figuring out how to completely seal off the bladder and mold so no leakage will occur where the air supply line meets the bladder. A couple ways to provide good sealing include using a sealing flange (either flat or tapered) to hold the end of the bladder securely at the mold opening, or to use o-rings and a sealing plug to capture the bladder opening and seal it off at an inlet port.

Sealing flanges can be created by first forming an air inlet hole with generously rounded edges. The end of the bladder is left to the outside of the mold through this inlet. A cap with an air fitting in it is fixed over the end of the bladder, pinching it against the mold to create a tight seal. If the bladder is made of a thin film material that won't allow it to stretch and form a good seal at the air inlet (such as with nylon or polyethylene films), sealant tape, or an o-ring or gasket can be used to fill any gaps in the film and better seal the bladder.

Capturing the bladder with o-rings at an inlet port can also produce a very stout seal. To create a good o-ring seal, a grooved metal or plastic plug is used to hold the o-rings in place. These o-rings then capture the bladder and force it against a sealing surface to make it air tight. The plug itself can be drilled and tapped to accept air line fittings for easy connection to the pressure source.

Inflatable Bladder Lamination Schedule

Resin System: High Viscosity Epoxy
2 Layers - Fine (3K) 2" Carbon Sleeve
2 Layers - Heavy (12K) 2" Carbon Sleeve

Materials and Supplies

Aluminum-filled Casting Epoxy
Delrin Plastic (3/4" Thick)
Plexiglass Sheet (3/16" Thick)
6061 Aluminum Rod (3" Diameter)
Latex Rubber Tubing (1" Diameter)
Mold Release/Parting Wax
Mixing Cups and Sticks
Spreader
Disposable Gloves
File or Rasp
Clean Rags/Cloths
Scissors
Double-sided Tape
Bandsaw
Sanders (Belt and Spindle Types)
Sewing Machine
Machine Lathe and Misc. Machine Tools
Clamps (C-Clamps and Pipe Clamps)
Air Fittings and Miscellaneous Fasteners
Shop Air Source (Up To 100 psi Pressure)

The inflatable bladder lamination schedule, along with materials and supplies, used in this demonstration.

A scrap piece of Delrin plastic was used to sculpt the pattern for this demonstration. Delrin can be a bit pricey, but it cuts, sands, and polishes nicely.

Draw the front and side profiles of the gear shift knob on paper, and cut them out for mounting on the plastic piece.

Cut the Delrin into two pieces of equal size, and use double-sided tape to adhere them together.

Spray adhesive will adhere the front and side profiles to the part.

Align the pieces well, and apply firm pressure so the tape will hold them together through the sculpting process.

Use a bandsaw to cut one of the profiles…

…and then rejoin the pieces together…

A belt sander can quickly remove material and shape the convex curves on the knob. Avoid creating undercuts at the parting line.

…to cut the other profile. The resulting part will have the proper shaping from these two directions.

Likewise a spindle sander can remove material from concave areas on the knob.

Mark the bottom profile (in this case, a simple circle) with a template. This will help guide the next sanding processes and minimize over-sanding.

Hand sanding can help develop any other areas that need further refinement.

Apply successive grits of polish to shine up the knob.

The final shift knob pattern should have a well-shined surface.

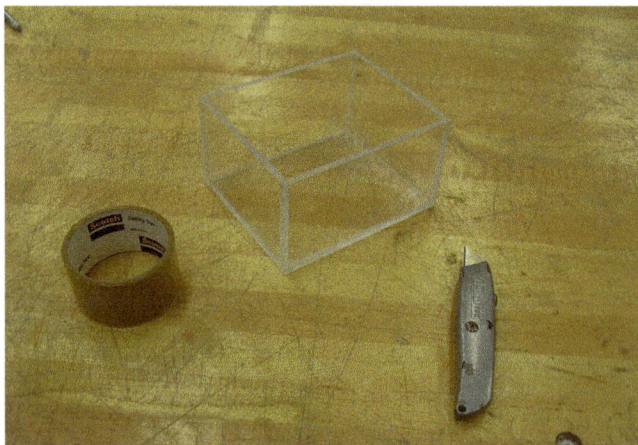

Create a box as a casting form and apply wax to it. Smooth acrylic makes a quick, effective box when sealed with clear packing tape.

The mold should be strong enough to not flex, fracture, or warp when the bladder is inflated in it at the desired pressure. This mold provision alone makes this technique of molding very tricky for general use, usually requiring a strong mold system made of machined metal or otherwise well-reinforced material. Pressure for bladders typically ranges from 20 psi to over 150 psi, depending on the consolidation and the amount of void content needed in the laminate. However, even a seemingly low 20 psi can impose high forces on a large mold - considerably more than most unreinforced mold systems may be able to handle. Considering these mold strength issues, it is wise to carefully test and monitor the pressure that is applied to any new mold system to ensure that the mold is not deforming in any way and that it can handle the pressures applied to it. It is also good to slowly inflate the bladder in the mold each time it is used. This will allow the bladder to expand more evenly and help keep areas of the bladder from inflating too quickly or rupturing.

There are few limitations on the type of reinforcement fibers that can be formed with a bladder system (though heavy fabrics and weaves with high crimp will require more pressure to completely flatten them against the mold faces), but the resin type should be carefully considered. As with any mold system, the bladder must be compatible with the resin and not chemically affected by it during layup and cure. The resin must also not be too thin and fluid, since the pressurized bladder may simply squeeze the resin out of the laminate and out of any gaps between the mold halves. Resins that are allowed to gel slightly before bladder inflation are also usable, though the timing, wet-out of the layup, and inflation can be a bit trickier to determine. For this reason, pre-preg composites tend to work best with bladders, but thick liquid resins will work as well.

One common measure of a liquid's viscosity is its "centipoise" (or "cps"). Water has a cps of 1, while motor oil (SAE 30) is 150-200 cps, and corn syrup is 2000-3000cps. Wet-layup resins work well when they are in the 300 to 500 cps range, while resins higher than 5,000 to 15,000

Carefully pry the two knob halves apart using a chisel.

Apply new double-sided tape to one half of the knob and press it into the box, with the bottom of the knob touching one side of the box.

Use double sided tape to adhere scrap pieces of plastic to the inside corners of the box. These will form small cavities that will align the mold halves.

Stir up the aluminum-filled epoxy using a power drill and a paint mixer attachment since the aluminum particles tend to settle over time.

Measure out the proper amount of resin for the casting...

cps tend to produce better results with inflatable bladders.

INFLATABLE BLADDER & MOLD FABRICATION TECHNIQUES

This chapter will demonstrate some inflatable bladder techniques through the creation of a hollow carbon fiber gear shift knob. These techniques will include the fabrication of a suitable mold "pattern", clamshell mold, and bladder sealing components, as well as special fabric processing and prototype mold clamping methods.

The "patterns" (or small plug forms) for the shift knob were made from solid, rigid plastic because of how well it releases from liquid casting materials, especially when it has been sufficiently polished and waxed. Delrin plastic was used for this part of the demonstration because its density and high temperature resistance allow it to be machined, cut, and sanded very cleanly. It also polishes up nicely and is rigid enough to avoid part deformation during various mold making processes. Therefore, simple woodshop tools can produce excellent results with minimal effort using this material.

Different techniques can be used to create clamshell molds, such as milling (shown in chapter 5) and composite tooling layup (shown in chapter 2) - though the latter requires

Continued on page 116

...and add the proper amount of hardener, as recommended by the manufacturer.

...and allow it to fully cure.

Use a drill and paint mixer attachment to thoroughly mix the epoxy components together.

Once cured remove the sides of the casting form...

Pour the filled epoxy into the forms...

...and then remove the pattern.

Sharp corners can be brittle with this casting material, so it is best to round them over using a utility knife for better mold longevity.

Wax all surfaces of the mold half (outside as well) at least three times using paste wax.

Apply new double-sided tape to the first knob half and place it back in the mold.

Align the second half of the pattern with the first half, and apply firm pressure to seal the two halves together.

Use the acrylic box sides as forms for the second half of the casting, sealing the acrylic with clear packing tape…

…and then using silicone sealant to seal the corners and edges of the mold.

Mix up more filled epoxy, and pour it into the forms and over the mold and pattern.

Once cured, carefully remove the casting forms…

…and use a sander to square up the sides of the mold. This will help when the molds are clamped together later.

Round-over the corners of the mold using a file or rasp.

Insert a wedge between the mold halves…

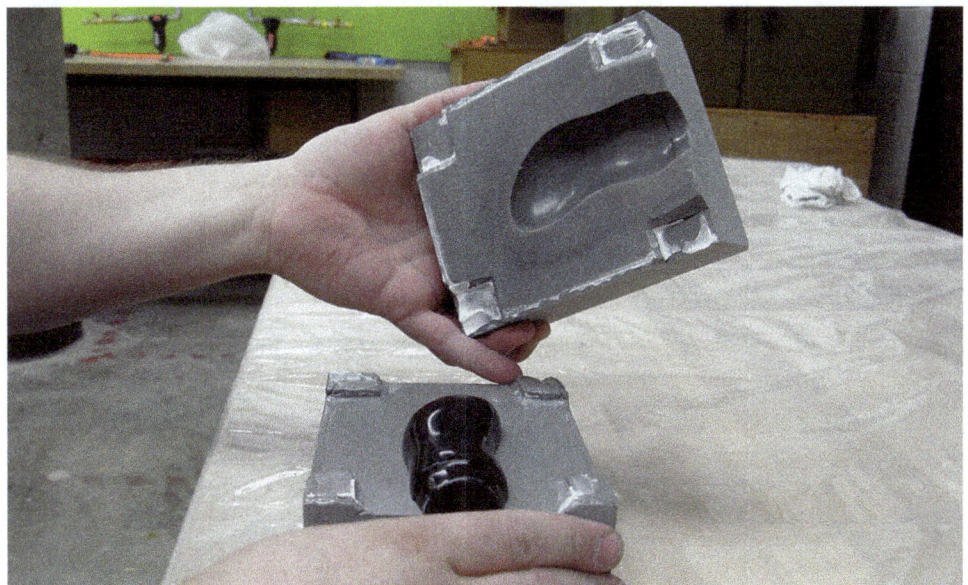

…and separate them to expose the pattern within. Remove the pattern and clean off any silicone sealant or other matter from the molds.

Thoroughly wax all surfaces of the mold prior to use.

significant build-up of material. But another method, demonstrated in this chapter, includes a two-step process of casting a metal-filled resin around a pattern to produce one side of the clamshell mold at a time. This specific method creates molds that fit perfectly with each other, but that also have some durability and strength limitations. Regardless, these molds can still be very effective for some uses, especially when they have been cast in thick sections or when backed up with reinforcing material. Therefore, this method of mold making works well for short-run parts at relatively low pressures, but high production volume and high pressure inevitably require the fabrication of molds machined from billet metal.

The casting material used in this demonstration is an aluminum-filled epoxy, called Freeman 801 Casting Epoxy, from www.freemansupply.com. This particular mold substrate will limit the heating capabilities of the mold (as its maximum temperature is around 140 degrees Fahrenheit) so the use of high-temp cure pre-preg is out of the question - although higher-temp filled resins are available for such applications. This material can still handle bladder inflation pressures well, if cast with thick enough walls, and it will provide good mold alignment if formed with proper mold registration built in.

As mentioned earlier, sealing a bladder can be an especially tricky feat, so special sealing components need to be machined for this purpose. For this demonstration, custom

A machine lathe was used to create the sealing flanges for the latex bladder. This particular flange setup was designed to be used with other molds as well.

The final pieces have matching beveled flanges that pinch the latex tight, to form a seal.

…and stretch the latex bladder over the flange to test for fit. This 1" diameter latex tubing was purchased from Piercan USA, Inc. (www.piercan.com).

Trace an alignment guideline on the mold where the sealing flange will meet it.

Bring the sealing flanges together…

Add any air fittings as needed…

…and snug them together with fasteners.

Wax all surfaces of the flanges to keep resin from permanently locking them together during inflation and cure.

...and turn the sock inside-out so the sewn seam is on the inside.

Sew up one end of the 3K carbon sock to create a nice seam at the end. This 2" diameter sock was purchased through www.sollercomposites.com.

Thickened epoxy was used for this layup to minimize "squeeze-out" of the resin (this 12,500 cps resin was purchased from www.uscomposites.com).

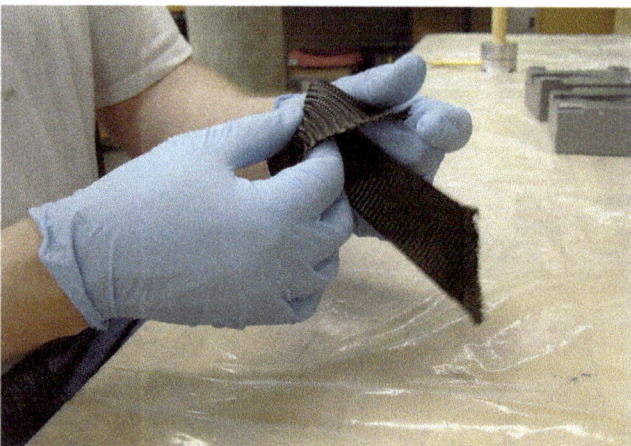

Carefully pull the open end of the sock back...

Use a spreader to press the thick resin through the sock. Apply resin to both sides for complete fiber saturation.

flanges were machined from scrap 6061 aluminum rod, and drilled and tapped to accept air fittings. For production molds, these components could be bolted onto the mold for ease of service, or even permanently machined into the mold itself. The flange components in this demonstration were developed as modular pieces that can be clamped to multiple molds, as needed, for different prototype parts.

To create a seamless, tubular part, carbon sleeving tends to work best, with fine weave sleeve for the outside layers and a heavier inner weave for fast build-up and strength on the inside. The 2" diameter sleeves in this demo were purchased from www.sollercomposites.com.

The open ends of sleeves can cause some aesthetic issues when they are laid up into the closed end of a mold (like this gear shift knob). But, because it is a fine-filament, flexible fabric, carbon fiber can be easily sewn with a sewing machine - so the end of the sleeve can be closed off quickly with a simple stitch. For other odd-shaped or enclosed forms, regular reinforcement fabric can be likewise sewn at its edges and laid into the mold with the bladder inserted in an opening at the seam.

Production molds that use inflatable bladders are usually "locked" together

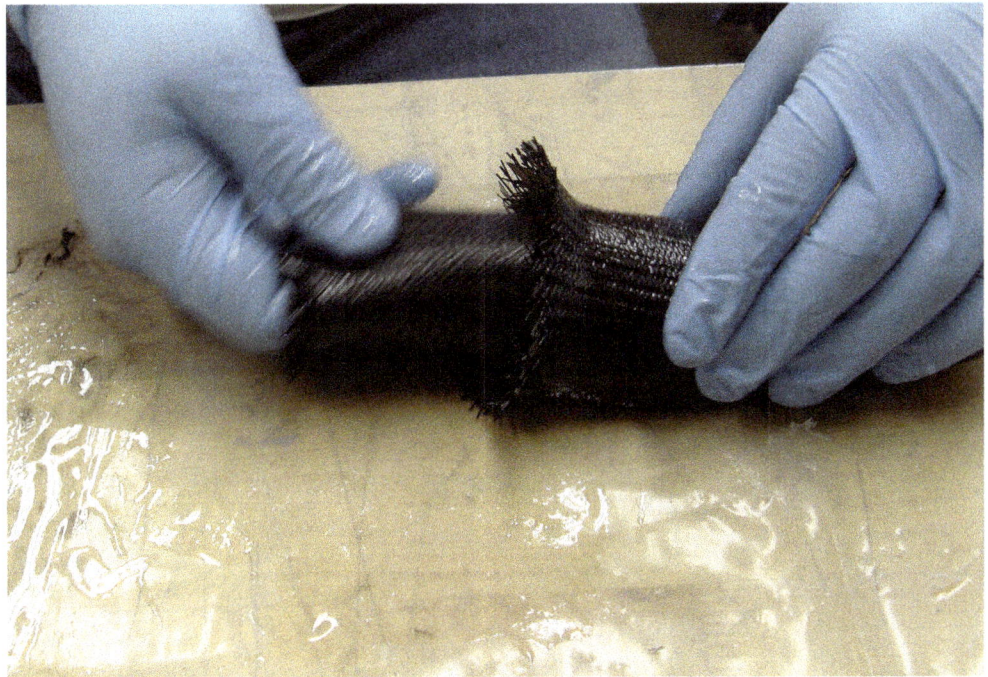

Insert the sock layers into each other...

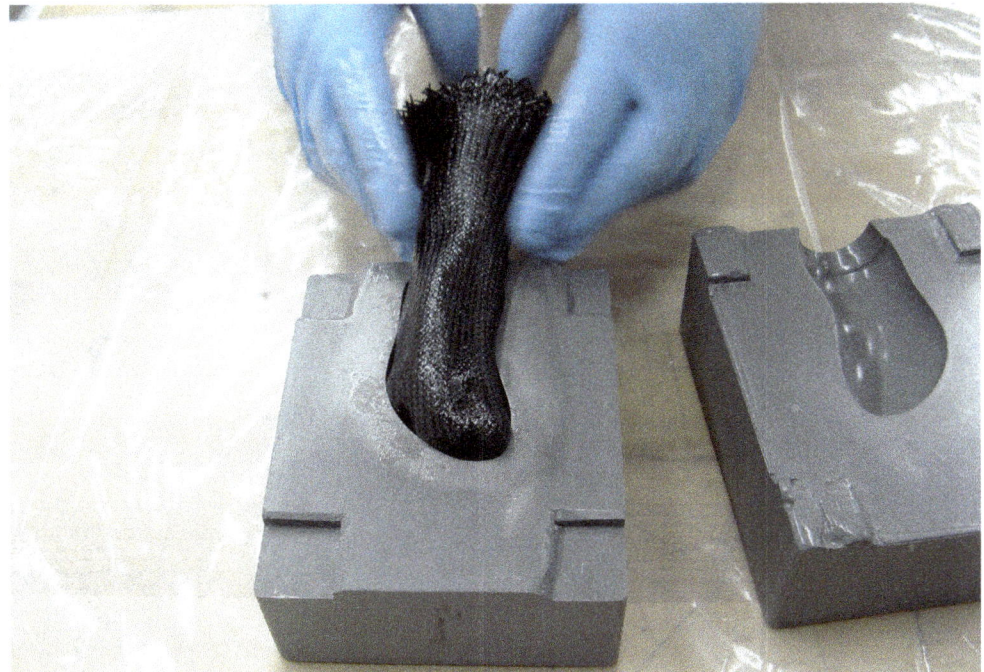

...and place them in the mold.

119

Trim off any excess sock...

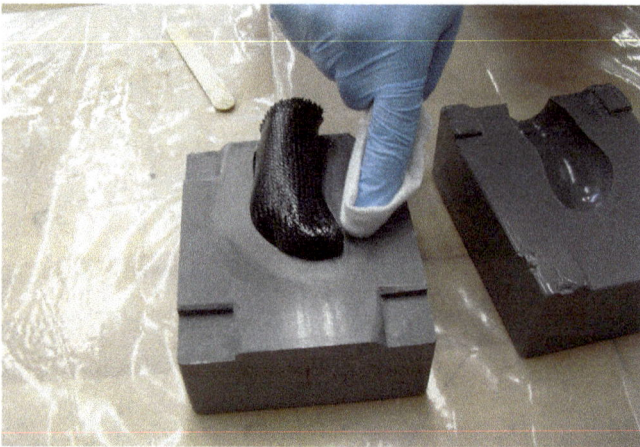

...and then wipe up any spilled resin that remains on the mold.

Insert the latex bladder into the cavity of the carbon sock and mold...

through mechanical fastening or hydraulic/pneumatic clamping methods. Thermoset prototype molds, such as the one in this demonstration, can be brittle in thin sections, so holes for bolt-through or threaded-in clamping hardware near edges of the material may cause fracturing issues. To avoid such problems, clamps or specially sized jigs secured around the prototype mold can effectively close the mold during bladder inflation and part curing.

Always test the pressure capabilities of a new mold prior to putting it into service. Thick, reinforced molds are rarely at risk for catastrophic failure at the typical pressures used with bladders (below 150 psi), especially when they are of small size or surface area. Usually, the biggest problems happen when a mold is improperly closed, a gap opens up, and the bladder over expands and ruptures. Large molds, though, can experience significant forces and can be especially prone to problems if not designed or handled correctly. Therefore, use good safety practices in testing out a new mold. Ensure secure attachment of clamping hardware while testing and using the mold. Also thoroughly check for any mold flexure or fracturing by incrementally increasing bladder pressure and carefully inspecting the mold until it is at its target molding pressure.

Wet layup in an inflatable bladder/cavity mold like this is generally the same as for other lamination methods. One caveat to bladder molding, though, is that the laminate from one half of the mold typically needs to bond with laminate in the other half, so a bit of overlap between the two laminates is necessary to connect the part halves. When pre-preg is used, the slightly stiffer properties of the pre-preg make bridging these halves manageable, especially when the molds are large enough for the fabricator to manually reach into the closed mold and apply the seam-bridging material. Wet reinforcements can be much more difficult to manage, however, because they tend to either slump into the mold cavity (away from the seam), or onto the mold flange (causing difficulties with mold closure). This can be mitigated somewhat by performing the layup, and then placing the partially inflated bladder in the mold to hold these edges in place while the top half of the mold is placed in position and secured. For tubular laminates, the natural structure of the woven sleeve reinforcement can significantly minimize these difficulties because it is already connected all the way around. However, sleeves tend to contract in length when inflated, so they should be laid into the mold as close to full inflation size as possible to maintain proper part geometry. After layup, securely close the mold and inflate the bladder, double-checking for fit.

A laminate's cure can be accelerated by placing the mold in an oven or by incorporating in-mold heating elements. Be sure to take the heat limits of the bladder into account and avoid exceeding the maximum service temperatures of both the bladder and the mold materials. If curing at normal room-temperature, a bladder-formed part will cure the same as for other room-temp laminates.

Cured parts can be demolded by completely releasing pressure from the bladder, unclamping the mold halves, removing the bladder, and extracting the part. Small parts can be removed by inserting a finger or long implement into the hole left by the bladder, and then by lifting the laminate out. For large parts, or those with small bladder openings, it may be helpful to design knockout-pins into

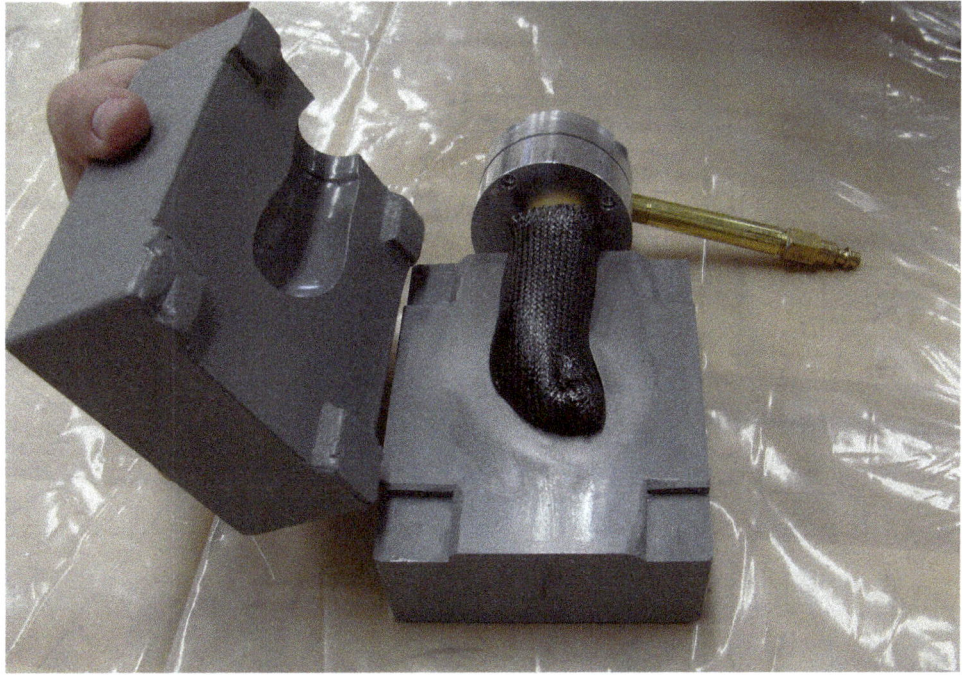

…and close the mold halves around the laminate. Avoid pinching the laminate.

Carefully apply clamps to the mold, applying snug (but not overly-tight) pressure.

121

Use pipe clamps or bar clamps to hold the sealing flanges in place...

the mold system to aid in part removal.

The first few parts from any inflatable bladder mold may have some imperfections in them. Take note of the location of voids or fabric separation and make adjustments in the layup, bladder sizing, or pressure to remedy these problems. Completed parts can then be trimmed and finished as usual.

CONCLUSIONS

For hollow, seamless composite parts, inflatable bladders can help produce exceptional results. Bladders can be made from films or rubbery materials, but should be impermeable to air and able to handle high inflation pressures inside a closed mold to work effectively. When sealed completely, the elevated pressures these bladders place on the composite laminate can significantly increase the finished surface quality while lowering the void content of the final parts.

...and apply regulated shop air to the mold. This mold was tested to over 100psi without problem. Always carefully test new molds prior to layup.

After the part has cured sufficiently, remove the shop air first. Next, remove the clamps, and then use a wedge to open the mold halves.

Slowly and carefully pull out the bladder to keep it from tearing. Bladders like these can be used multiple times if carefully handled.

The final part should have excellent consolidation and low void content. It is now ready for finishing and bonding with a threaded insert for installation into a vehicle.

Chapter Seven

Resin Transfer Molding Techniques

Optimized Parts, With Less Resin Mess

INTRODUCTION

When it comes to creating production parts with excellent fiber-to-resin content, one of the least expensive and most repeatable methods is through a process called "resin transfer" molding. With resin transfer, liquid resin is moved from a supply source (such as a supply container) into a "preform" (such as dry fabric layers placed in a mold) by the application of pressure (such as atmospheric pressure or a piston system). This chapter will demonstrate some of the techniques used to carry out a vacuum-assisted version of resin transfer that, with practice, is as easy to perform as typical vacuum bagging techniques.

Optimum resin-to-fiber ratios can be controlled in a composite by introducing the "perfect" amount of resin into the reinforcement through resin transfer methods - as with the concept car body panel, demonstrated in this chapter.

RESIN TRANSFER BASICS

The general term "resin transfer molding" applies to a wide range of processes through which resin is forced into reinforcements within a mold. A very common version of resin transfer is called "resin infusion", which also goes by the names of "vacuum infusion process" (VIP) and "vacuum-assisted resin transfer molding" (VARTM). With this type of resin transfer, a vacuum draws resin into dry reinforcement fabric in the mold. If done properly, this resin transfer technique will produce an optimized composite with only that resin necessary to hold the fibers together, producing a naturally low void content and a near-perfect fiber-to-resin ratio composite. Rather than putting excess resin into a fabric and then sucking it out (as with vacuum bagging) or relying on a manufacturer to produce expensive pre-preg for you, this process allows only the right amount of resin to enter the fabric to begin with.

When performed as a vacuum-assisted procedure, resin transfer starts when a vacuum is applied over dry

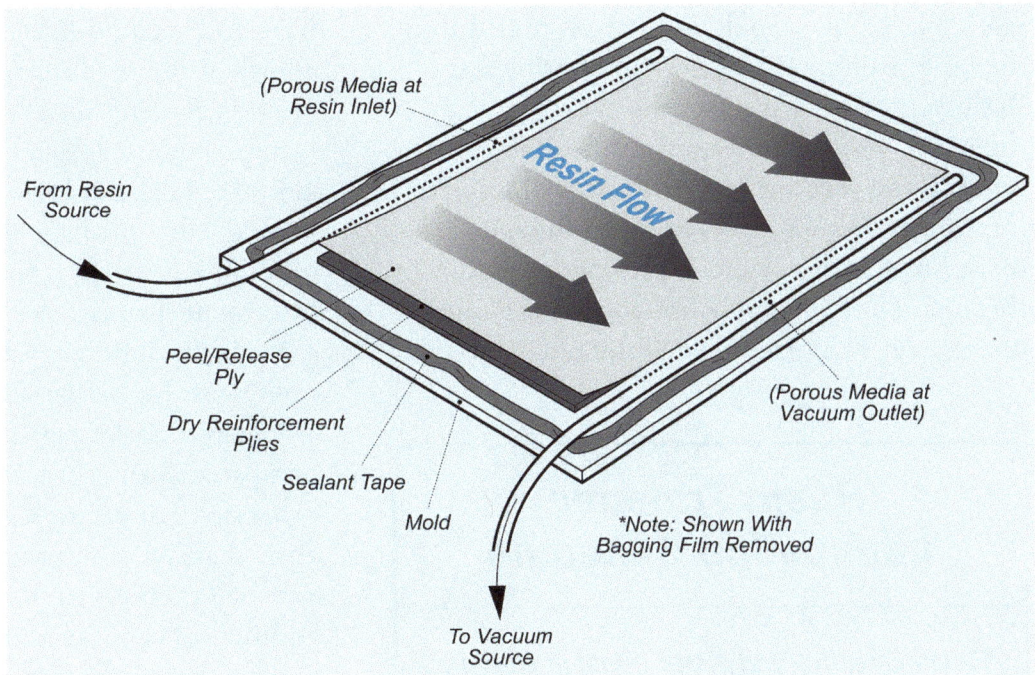

Resin transfer (or "resin infusion") works by drawing resin into the laminate by a vacuum.

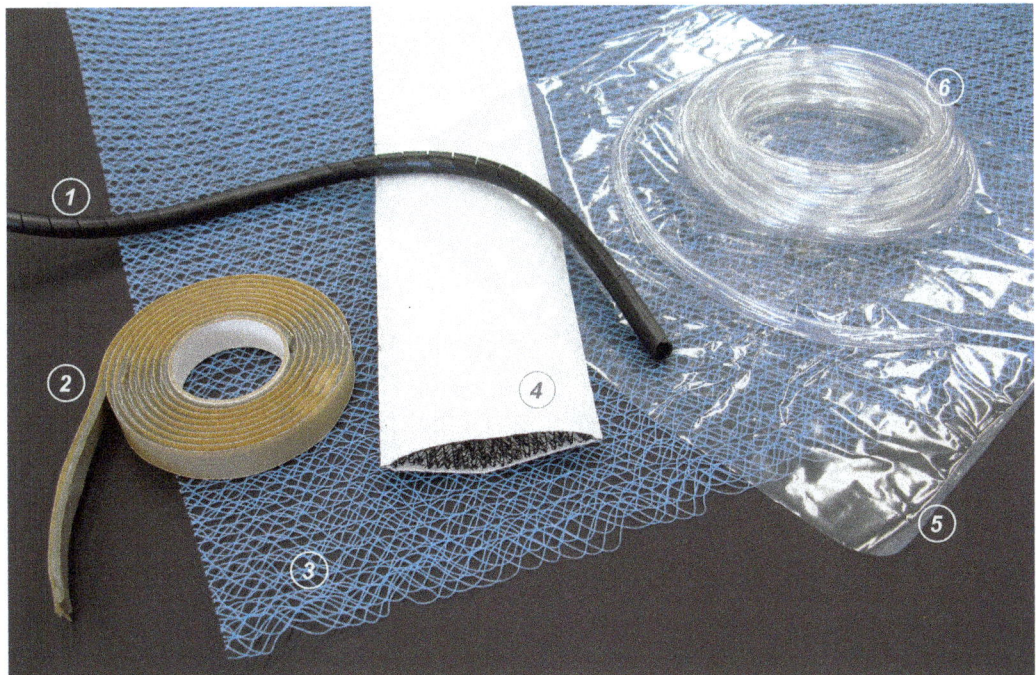

Resin transfer requires several unique materials: 1) Spiral tubing , 2) Sealant tape, 3) Nylon matting flow media (for large parts), 4) Filter jacket, 5) Bagging film, and 6) Vacuum tubing.

fabric placed in a mold, and resin is pulled into the mold from a mixing container or bucket. Resin supply lines connect the resin source to the mold, and vacuum supply lines connect the mold to the vacuum source. Resin will migrate into the evacuated spaces between the reinforcement fibers as atmospheric pressure pushes it through the resin supply line toward the vacuum source. The placement of the resin supply

Resin Transfer Lamination Schedule

Resin System: Polyester Infusion Resin
2 Layers - 5.7 oz. Carbon (3K - 2x2 Twill)
4 Layers - 12 oz. Fiberglass (Double Bias)

Materials and Supplies

Spiral-cut Polyethelene Tubing
Vinyl Tubing (3/16" ID)
Release Ply Fabric
Bagging Film
Enkafusion Infusion Jacket
Mold Release/Parting Wax
Mixing Cups and Sticks
Spreader
Disposable Gloves
Clean Rags/Cloths
Scissors
Clamps (Squeeze-type)
Masking Tape
Sealant Tape
Vacuum Source
Resin Trap

The resin transfer lamination schedule, along with materials and supplies, used in this demonstration.

lines and vacuum lines, along with the permeability of the reinforcement fiber and resin viscosity, will determine how quickly the laminate is impregnated. Efficient resin flow between the laminate plies can be further facilitated by adding "flow media" - which usually consists of randomly oriented strands of plastic or reinforcement material that provides additional space in the laminate so resin can quickly migrate in for better laminate saturation.

Resin transfer has numerous benefits when compared with other layup and molding processes. For example, it requires less consumables than vacuum bagging, meaning that each part will cost less to produce, even when the addition of resin/vacuum lines and flow media are factored in. In fact, many typical costs are offset by the simplicity of this process because it alleviates the use of brushes, rollers, protective clothing, respirators, and some skilled labor costs. In certain production situations, resin transfer can even eliminate the need for costly ventilation equipment because there is little or no contact with liquid resin - minimizing the health risks related to chemical contact and VOC exposure.

Resin transfer can be especially beneficial in large production settings. Parts produced through resin transfer have much more uniformity in their weight and strength compared with common wet layup methods. These processes have excellent repeatability so that the quality and properties of the resulting laminates are very close to each other and predictable. This fact alone makes resin transfer great for production runs, but its inherently high part production speed (especially with heated molds) greatly improves the production benefits of this process. Furthermore, the performance between completed parts is nearly the same as pre-preg parts - and with a similar resin-to-reinforcement content - though it does not typically have the same compaction against the mold surfaces as parts processed in an autoclave (as pre-preg commonly is). These benefits over pre-preg are magnified when resin transfer's low cost is compared to pre-preg.

Even though it is considered an "advanced" process, resin transfer requires less skill to perform than typical layups, especially once the process has been "dialed in" and perfected for a particular part. Typically, a lack of skill involved in impregnating a fabric with resin accounts for a large degree of error and variability in a composite. So supplanting this step with resin transfer all but alleviates many of the associated resin-related part quality issues.

Because it is so similar to vacuum bagging processes, those who are experienced in bagging parts will often find that vacuum-assisted resin transfer is a natural progression for their skills - even a sideways step to an easier process. With resin transfer, set up time is unlimited (rather than fighting against the clock as you watch your resin harden in the cup), there is considerably less mess (which means less time bent

Prepare and wax the mold, just as you would with wet layup methods.

Spray a small amount of spray adhesive onto the mold surface to stick the reinforcement plies to the mold.

Place the first layer of reinforcement in the mold…

over the sink with a handful of Fast Orange cleaner), chemical "odor" is minimized (which always pleases the significant other), and cost is reduced (leaving more greenbacks for that "next great project").

With so much going for it, resin transfer may sound like the perfect solution to several composites fabrication problems, but it does have some shortcomings. As attractive as it seems, resin transfer has its own set of quirks - just like that cheerleader in your high school biology class. For one, resin flow is extremely important and will ultimately make or break a part. Secondly, surface quality can be a problem, especially when surface aesthetics are critical. Additionally, cores, inserts, and molded-in fasteners can cause big headaches if not processed correctly.

Without good resin flow, a resin infused part will simply not work. So a properly selected resin and fabric can make a big difference in bringing that important composite part out of the mold with the quality it needs.

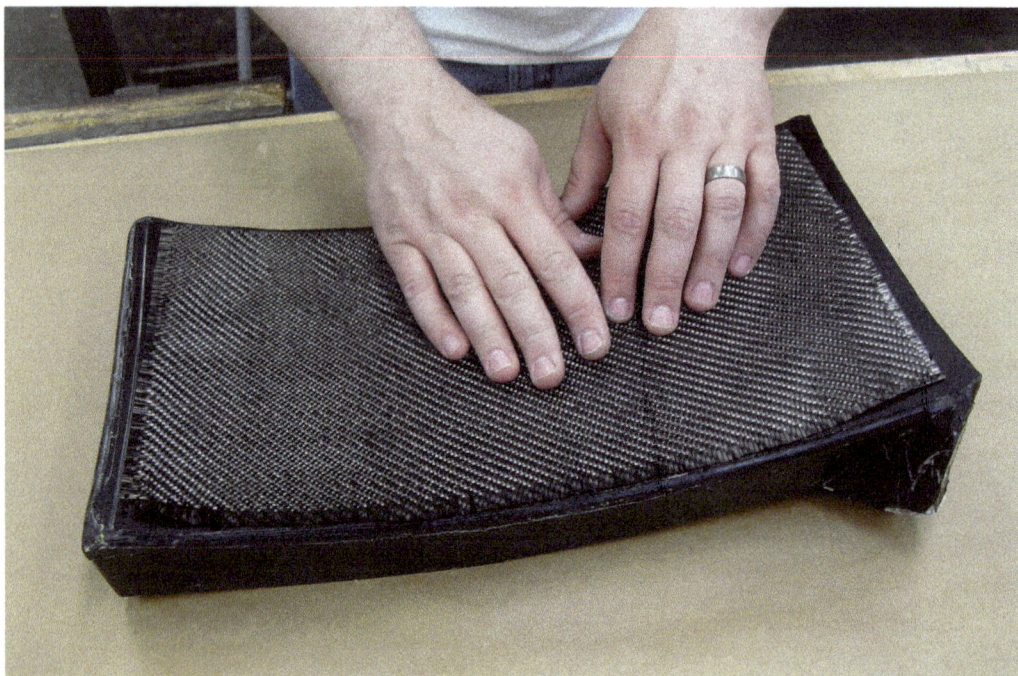

…and smooth it from the center out to remove any wrinkles.

Resin viscosity can significantly speed up or slow down resin transfer, but the weave of a fabric can do the same. In this regard, it is best to help these two components to "play nice" with each other.

While thick resin works best for inflatable bladder processes, thin resins tend to work best with resin transfer molding because of their ability to permeate the fabric quickly. Thick resin may be difficult or impossible to use with resin transfer since it simply will not move through the fabric completely before it cures. As a general rule, resins below 200 cps tend to give the best resin transfer results, though thicker resins may still work if they are slower-cure types and are given enough time to infuse. Regardless, the thinner the resin, the better it will infuse.

Vinylester has been a long-time favorite for resin transfer because of its comparatively low viscosity and good physical properties. But, understanding the importance of thin resins to the composites industry,

For successive layers, apply sparing amounts of spray adhesive to the previous layer, and then add more reinforcement, as needed.

Once the lamination schedule is complete, place a layer of peel ply over the dry reinforcements.

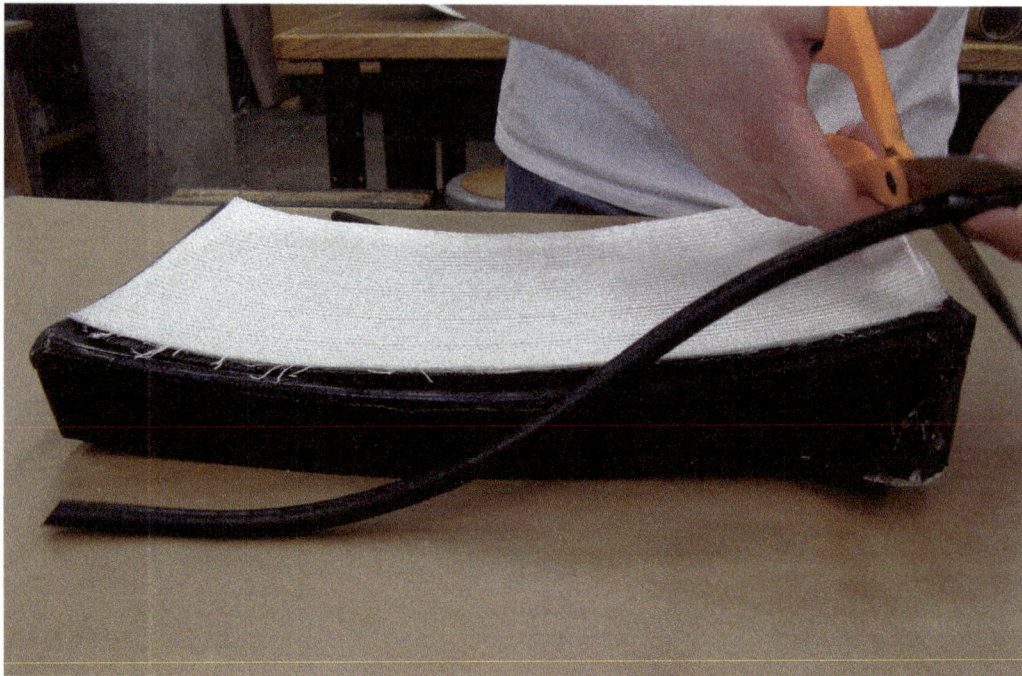

Measure and cut the spiral tubing (for the vacuum lines)…

…and then measure and cut the EnkaFusion filter jacket (which will be attached to the resin supply line).

some resin manufacturers have started to develop other infusion-specific options outside of vinylester. For example, the Mas Epoxy corporation (www.masepoxies .com) now distributes a 200 cps epoxy, and Cook Composites and Polymers (www.ccponline.com) produces a similarly thin polyester resin.

In addition to resin viscosity, proper fabric selection is important to good resin flow. Many fabricators have dabbled in resin transfer processes with common and inexpensive plain weave fabrics only to be disappointed by the poor resin flow and overall high porosity of their final infused parts. The frustration that ensues from these ventures seems to stem more from a lack of experience than from resin transfer's effectiveness as a valid process. With infusion, resin flows along the spaces between the actual fibers - so the less crimp in the fabric weave, the faster the resin flow. Therefore, loose weaves (such as satin, leno, basket, and twill) and knit fabrics tend to be most effective with resin transfer processes.

Another frustration that fabricators find is in optimizing the resin transfer process for their specific parts. Generally, setup can be time consuming and entails running through several test parts until the proper placement of vacuum and resin lines is determined - especially for large or complex parts. This can quickly translate to significant cost and headaches unless less expensive fabrics (of the same weave and thickness) are used for testing. To troubleshoot and perfect a flow problem, "infusion maps" can be helpful. After the vacuum has been applied and the resin flow open for a given amount of time, the flow can be stopped (with a clamp on the resin line), and the shape of the infused resin traced on the vacuum bag with a permanent marker. After it has been marked, the resin is allowed to flow again for the same amount of time, and the process is repeated until a topographical-style map is completed that shows the progressive flow of the resin across the laminate.

Results from an infusion map can enlighten the fabricator about where to

Wrap the spiral tubing with peel ply and hold it together with masking tape or sealant tape.

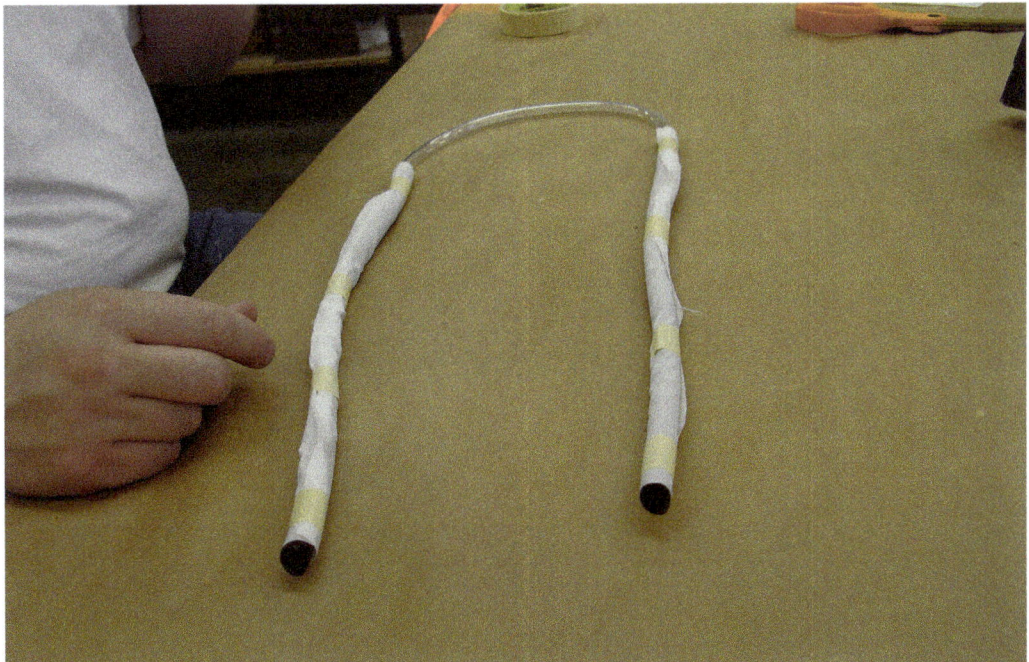

Use clear vinyl tubing to connect the vacuum lines to each other.

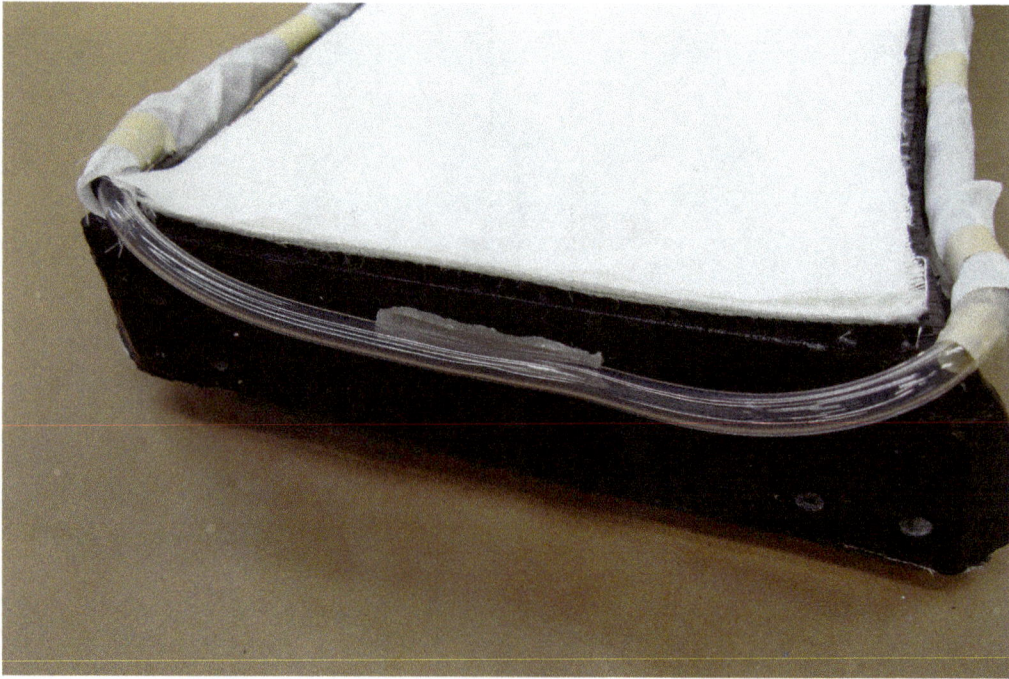

Adhere the vacuum tubing assembly to the mold using sealant tape.

place resin and vacuum lines for more efficient flow and resin use, and may even highlight the need for additional flow media (like nylon matting) as well as where to locate it on the laminate. Evenly spaced lines through-out the map will show consistent flow in the laminate, whereas localized areas of very close lines will denote poor flow. Adding sections of flow media in these slow-moving areas can significantly speed up resin migration. If the map shows that the laminate has consistent, even flow, but that resin is reaching the vacuum lines faster in some areas than others, further space out the resin lines from the vacuum line in those particular areas of the laminate. While this latter problem will not necessarily cause many problems with the infusion process, it can save resin from being wasted over the long run.

Surface quality can be a problem with resin transfer simply because laminate compaction is limited by atmospheric pressure. Infused parts commonly have the same surface voids that are evident on vacuum-

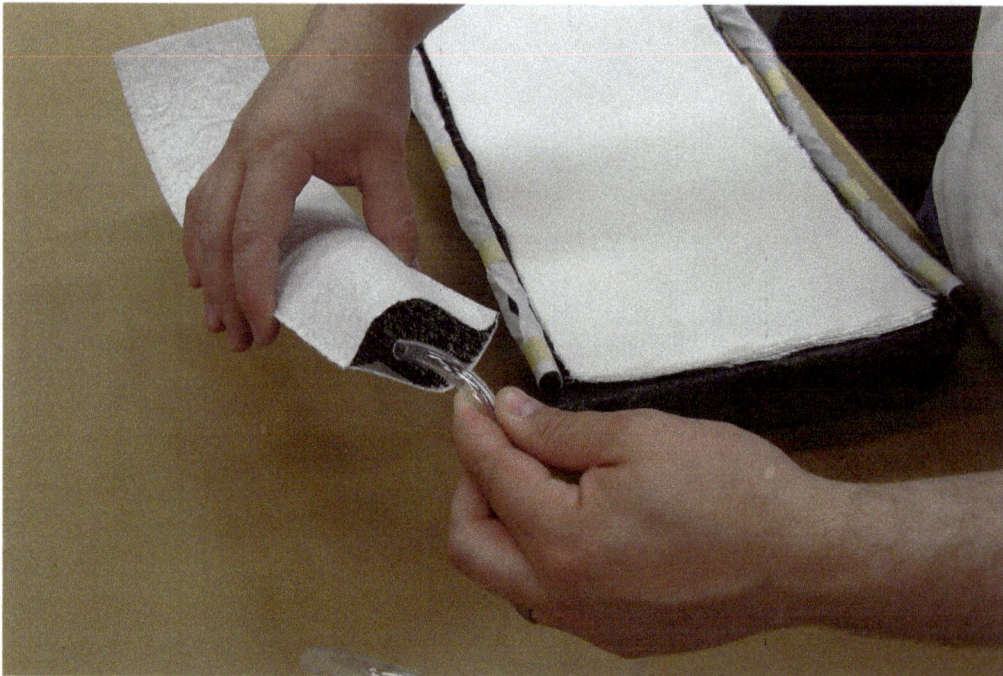

Insert vinyl tubing into the top side of the filter jacket until the end of it is halfway down the jacket. This will allow resin to flow evenly from the center.

bagged parts. To remedy this, gel coats and surface coats can still be used - although they will add weight to the part. Using unidirectional fabric for the first layer against the mold face can also help. Some fabricators have also seen success in first infusing a part, and then placing it under pressure (with compression molds, inflatable bladders, or an autoclave) to improve the surface quality.

With resin infusion, core selection is limited, and open-celled cores (like honeycomb) can be nearly impossible to use. However, some foam cores are made specifically for resin transfer processes (like Divinymat). Such cores facilitate resin flow between the laminate and core interface through small channels that have been scored into them, or by means of scrim cloth or other flow media that has been bonded to them. Some honeycomb-like cores are available (such as Lantor Soric), though true open-celled honeycomb will simply fill with resin and create an overly heavy composite.

Sealant tape can help adhere this assembly to the mold, as well.

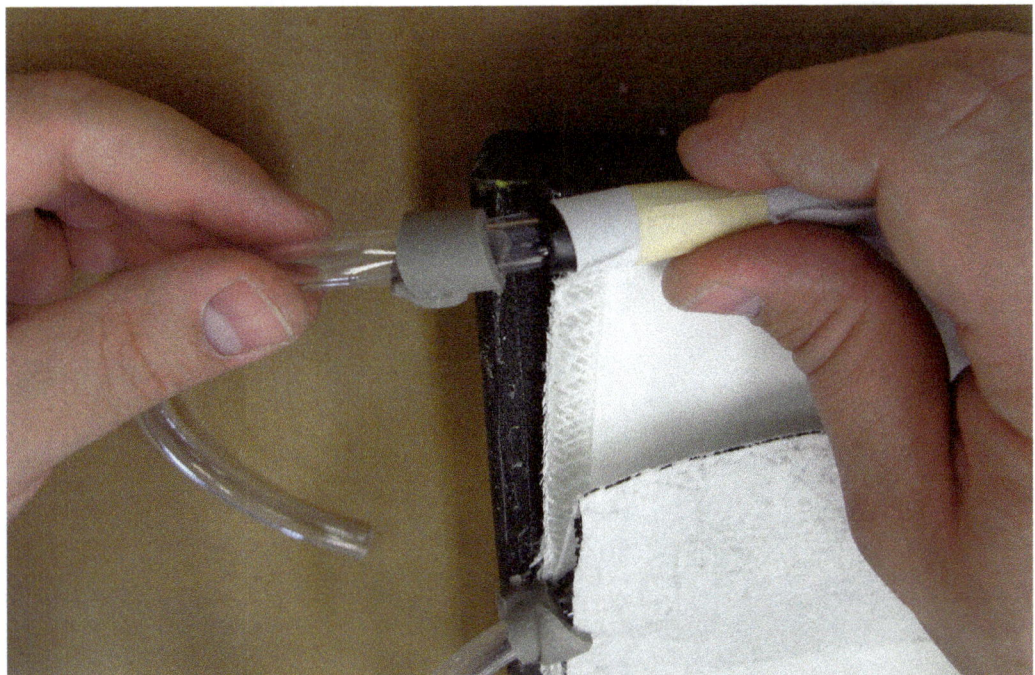

Use additional tubing to extend the vacuum line outside the edge of the mold.

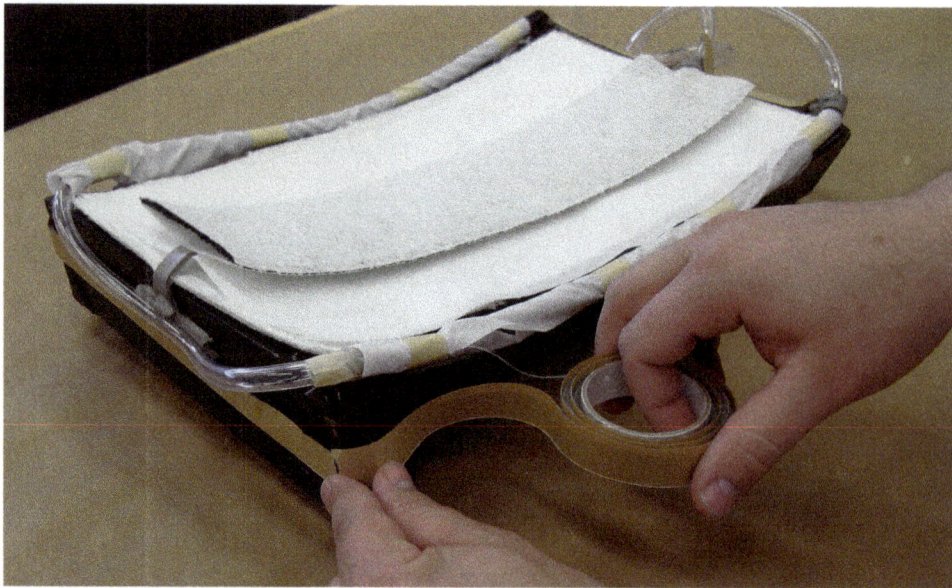

Apply sealant tape to the outer edge of the mold…

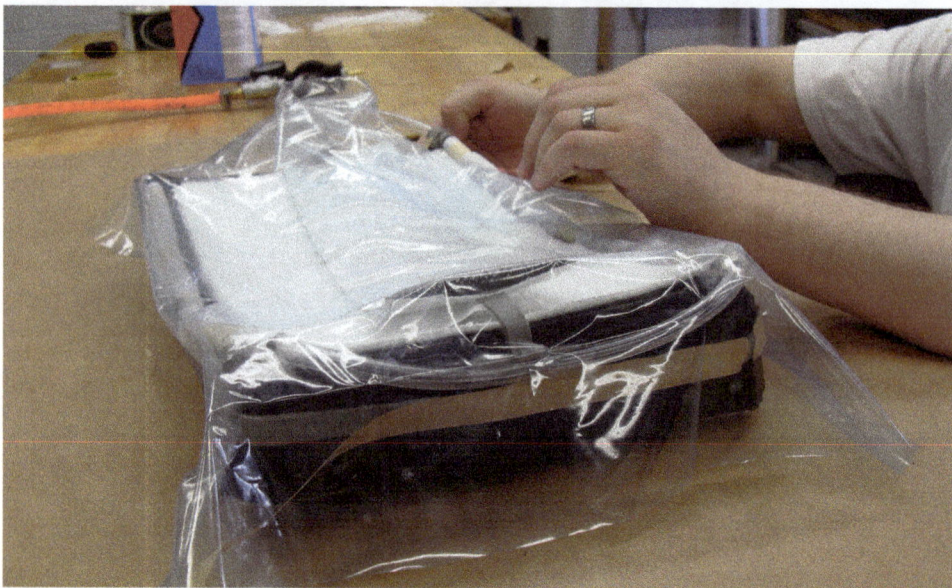

…and adhere bagging film to the entire mold (as is common with vacuum bagging techniques).

Create pleats and tucks where needed with additional sealant tape.

A couple of final considerations with resin transfer include the part geometries and molds used with this process. Infusion works most effectively on "open shapes" that are easily accessible from one side. "Closed shapes", like hollow forms, are still possible but can be extremely difficult to process.

Molds used with resin transfer should have the same robustness and air-tight quality required for any vacuum bagging process. Unlike inflatable bladder and compression molds, a resin infusion mold will usually only need enough strength to handle atmospheric pressure acting on it. These molds must have a good seal and be free of cracks, pores, or other imperfections that would compromise the mold's ability to hold a vacuum - all guidelines that should be followed for any mold that will be used to form a composite laminate.

RESIN TRANSFER TECHNIQUES

With resin transfer processes, molds are prepared as usual using paste wax and PVA, or water-based releases.

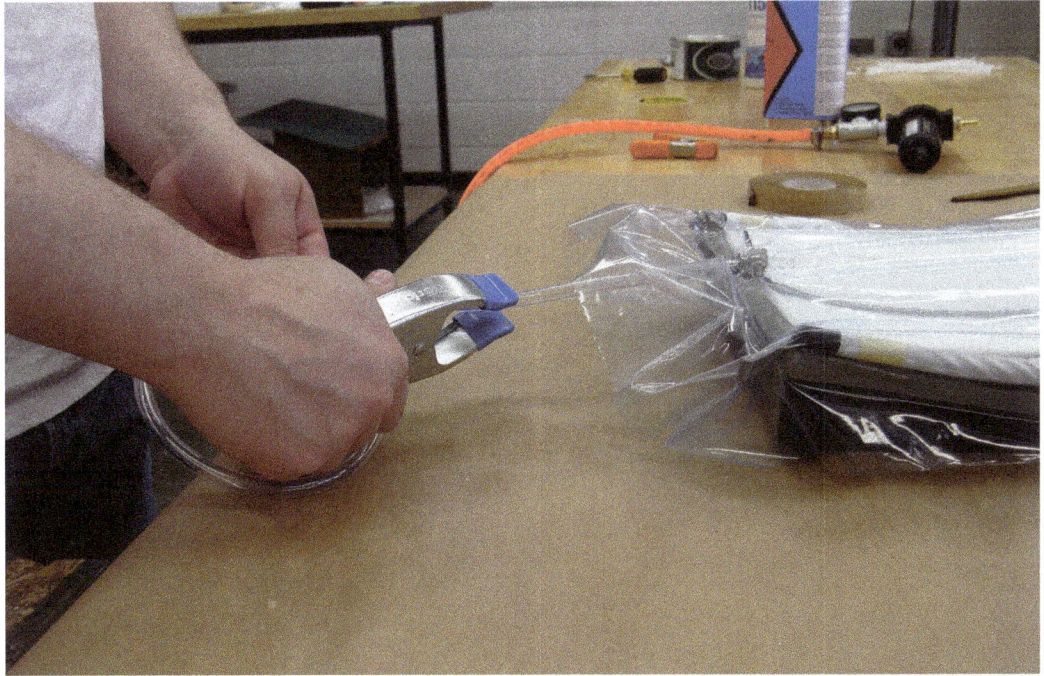

Use a clamp to close off the resin supply line.

Connect the vacuum line to the resin trap and vacuum source and check for any leaks. It is very important that the bagged mold has a tight seal.

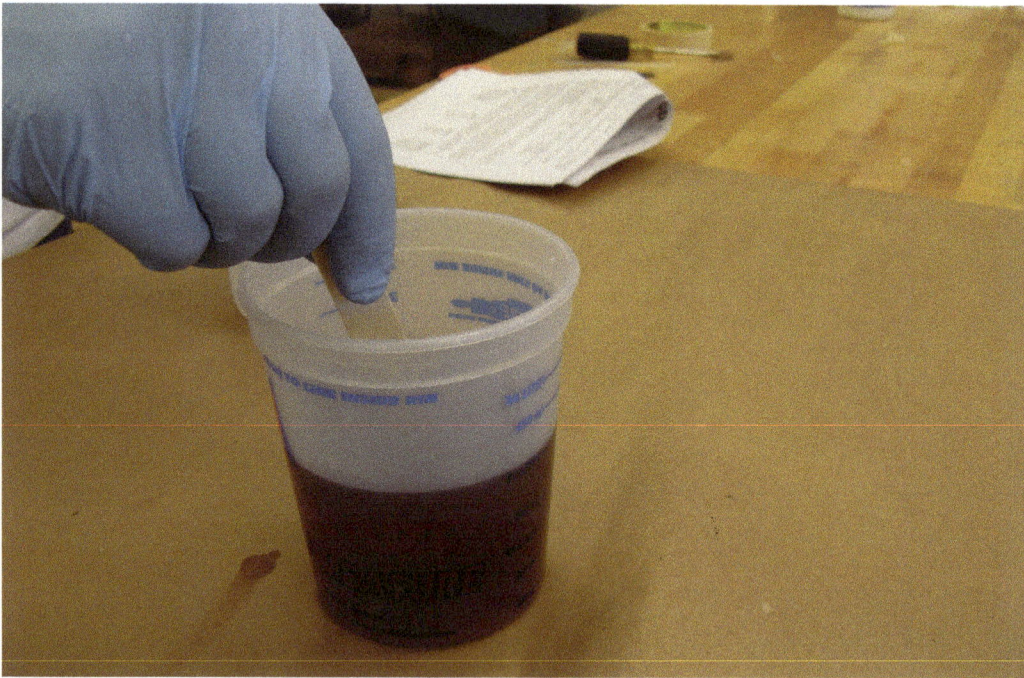

Mix up the resin, as prescribed by the manufacturer.

Reinforcements can be laid directly into the mold, or adhered to the mold walls using moderate amounts of spray adhesive. Additional reinforcement plies can be placed into the mold, using spray adhesive between them to hold the dry reinforcements onto mold features. Always place peel/release ply on top of the laminate prior to adding vacuum and resin lines or flow media - it will keep these media from permanently adhering to the laminate.

Vacuum lines and resin lines should be placed in the mold where they will be able to draw resin into the layers most efficiently and thoroughly. The placement of these lines will be largely determined by the geometry of the composite part, but vacuum lines are typically placed at the outside edge of the part. Small or square parts (less than a couple feet in width) can be set up with a simple resin line on one edge of the reinforcement stack, and a vacuum line on the other. For this small demonstration part, such an arrangement would have been sufficient, but an infusion jacket

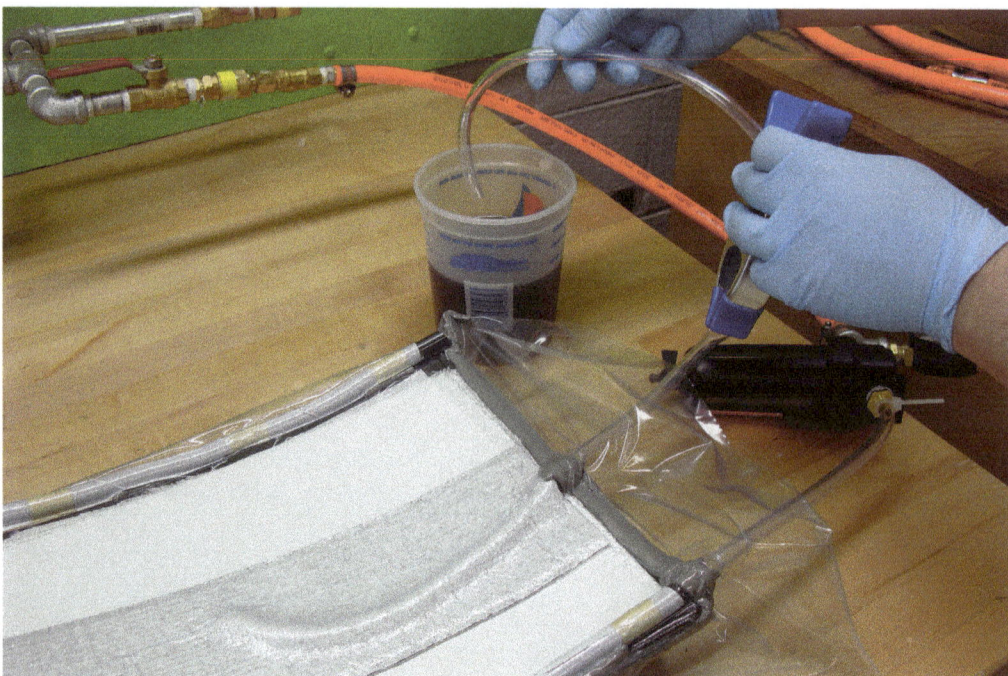

Insert the end of the resin supply line into the tub of mixed resin...

was also added to show how this process could apply to larger laminates as well.

Larger laminates, or those with complex shapes, will greatly benefit from infusion jackets and flow media. Infusion jackets can be placed at the center of a large part to shorten the distance between the resin line and the vacuum line. A resin supply line can then be directly inserted into the center of the infusion jacket, or a fitting can be placed into the center of the infusion jacket for connection to a resin supply line through the bag itself. Flow media, if needed, is simply placed alongside the infusion jacket over the peel/release ply in any areas where increased resin flow is needed. Connect sufficient tubing to both the vacuum lines and resin lines attached near the reinforcements.

After the resin and vacuum lines have been placed correctly, use common sealant tape and bagging film to encapsulate the whole system, similar to vacuum bagging processes, adding tucks and pleats as necessary. A

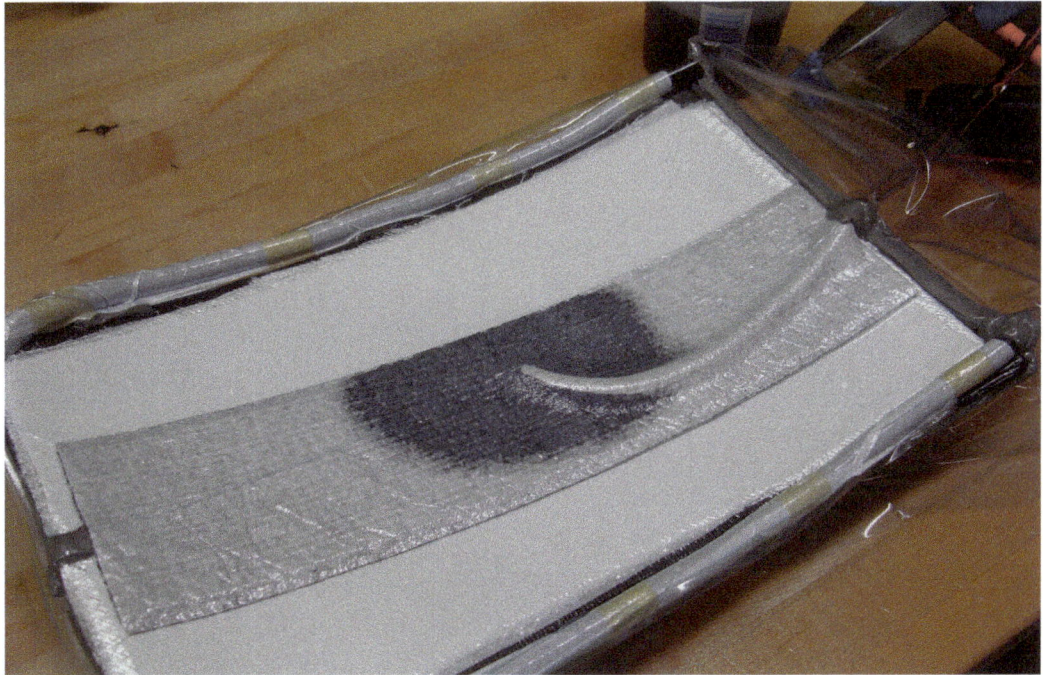

…and release the clamp on the resin supply line. Resin should immediately begin to flow into the jacket.

Resin should quickly infiltrate the jacket…

...migrate into the laminate plies...

complete seal is very important, so firmly apply the sealant tape to prevent leaks from developing at the edge of the bag.

A resin trap is imperative with resin transfer because resin will typically flow out of the laminate as some areas of the composite are thoroughly infiltrated more quickly than others. This tendency for resin overflow will be magnified if small leaks are present in the seal around the part, as air can enter the bag and push resin out the vacuum line. Small parts may be fine with relatively small resin traps, but large parts may require large ones (or even multiple traps run in parallel). Large resin traps are commonly equipped with removable tops, so the inside of the trap can be waxed, allowing excess, unused resin trapped in them to be easily removed.

Mix up the resin and open up resin flow only after the part has been evacuated and all leaks have been closed. Resin can flow very quickly (if done correctly), so be prepared to shut off the resin flow once the

...and move toward the vacuum lines. Allow this to continue until full fabric saturation is complete. This particular part took less than a minute to completely infuse.

part is thoroughly infused.

Once infusion is complete and resin flow is stopped, allow the laminate to cure under vacuum. This will maintain pressure on the laminate and facilitate consolidation of the fibers during cure. After it has cured, the part may be demolded, trimmed, finished, and put into service!

CONCLUSIONS

Resin transfer offers several benefits over other composite laminate optimization methods, but comes with its own set of limitations as well. Selecting the right materials for proper resin flow - and using a little patience and perseverance with this trial-and-error process - the average composites fabricator should be able to produce parts that have the same performance characteristics of pre-preg laminates, at a much lower cost.

Once the resin transfer is complete, clamp off the resin supply line and allow the laminate to cure completely under vacuum.

After cure, remove all bagging materials from the part...

...and remove the flow media and peel ply from the laminate.

Use a wedge to release the part from the mold.

The final, optimized laminate is ready for trimming and finishing!

Conclusion

One of my greatest joys as an industrial design instructor (who also happens to be a skilled composites designer and fabricator) is the opportunity to enlighten students about many of the design possibilities inherent in several different materials and manufacturing processes. In my encounters with bright (but often timid) new students, I have met many creative souls who dream of fabricating a wide range of intriguing, fantastic, and innovative products—but who often find their dreams clouded or stifled by their own inability to bring their designs into reality. At the end of their educational careers, however, it is inspiring to see how each student's capabilities have blossomed over just a few semesters and empowered them to turn their dreams into reality. It was with this end in mind that I wrote this book for both students and general fabricators alike.

Composite Materials: Fabrication Handbook #1 was originally written to educate readers on some of the more basic techniques employed to make simple composite laminates. This book—*Composite Materials: Fabrication Handbook #2*—was developed as a companion to *Handbook #1* to help clarify several of the more advanced techniques utilized by fabricators and manufacturers in the composites industry. As many fabricators have come to realize, composites offer a great deal of design flexibility when compared with traditional materials. However, they also require a new set of skills for those unfamiliar with them. Once these skills are set in motion, though, composite fabrication can truly become a fun exercise—especially when that first good component comes out of a new mold, ready for finishing and final use.

It is my sincere hope that readers will feel at least a bit more informed about how to build their own advanced laminate creations in a small shop environment as they utilize (or creatively adapt) the varied, practical laminate construction techniques explained throughout this book for their own projects. For those who, even after reading this book, may still be locked in indecision about whether or not to use composites for their next project, I can only echo the wise, immortal words of Paul Teutul, Sr. (from American Choppers fame): "Don't THINK; just DO!"

Action makes all the difference.

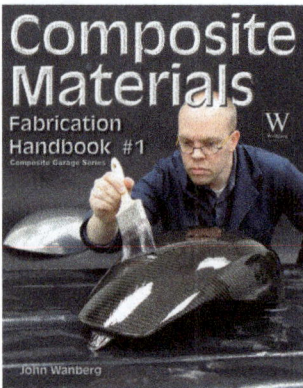

COMPOSITE MATERIALS FABRICATION HANDBOOK #1

Beginning Composites presents practical, hands-on information about these versatile materials. From explanations of what a composite is, to demonstrations on how to actually utilize them in various projects, this book provides a simple, concise perspective on molding and finishing techniques to empower even the most apprehensive beginner.

Topics include: What is a composite, why use composites, general composite types and where composites are typically used. Composite Materials Fabrication Handbook includes shop set up, design and a number of hands-on start-to-finish projects documented with abundant photographs.

Eight Chapters   144 Pages $27.95   Over 400 photos, 100% color

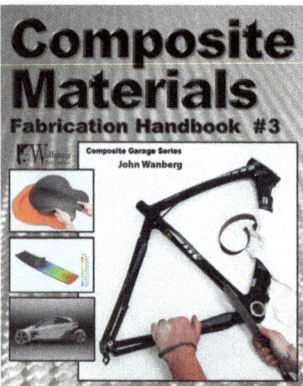

COMPOSITE MATERIALS FABRICATION HANDBOOK #3

Composite Fabrication Handbook #3 continues the practical, hands-on series on composites with helpful how-to projects that cover a variety of topics geared toward assisting home-builders in completing their composite projects. Handbook #3 expands further on mold-making techniques showing some special methods for creating molds and composite copies of existing parts, fabricating molds from clay models, and making advanced

mold systems using computer modeling software and CNC equipment. Several alternative methods of fabricating one-off parts are presented in this book, including molding over frameworks and human forms, as well as using stock composites to build simple structures. Composite repairs are also covered in this book, along with a primer on computer-aided analysis of composites structures and how professional fabricators build composites.

Nine Chapters   144 Pages $27.95   Over 350 photos, 100% color

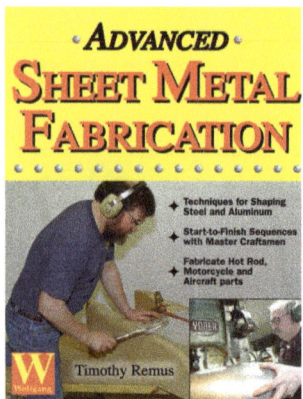

ADVANCED SHEET METAL FABRICATION

Advanced Sheet Metal Fabrication is a photo-intensive how-to book. See Craig Naff build a Rolls Royce fender, Rob Roehl create a motorcycle gas tank, Ron Covell form part of a quarter midget body and Fay Butler shape a aircraft wheel fairing. Methods and tools include

English wheel, power hammer, shrinkers and stretchers, and of course the hammer and dolly. Learn form the experts how to weld the various panels into one complete, finished piece with minimal warpage – with either TIG or gas.

Eleven Chapters   144 Pages $27.95   Over 400 photos, 100% color

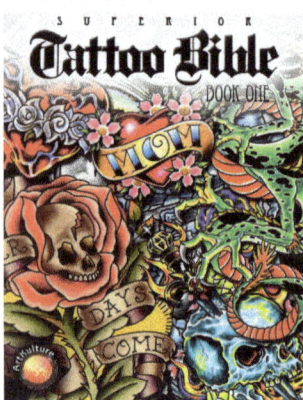

TATTOO BIBLE BOOK ONE

Whether you are preparing for your first tattoo or your twenty-seventh, you need artwork and designs that are just-right. Tattoo Bible, authored by Superior Tattoo, provides well over 500 pieces of unique flash art - flash never before compiled into one single book.

While most tattoo books available today concentrate on one specific genre, this Tattoo Bible covers many different genres and the ideas are end-

less. This is not just a book to add to your collection - this is your collection. You can combine different pieces of art from within the book, or just take them as is. This book is for you and your imagination to do with as you wish.

Published by ArtKulture, an imprint of Wolfgang Publications, with images that are both striking and very useful to both the tattoo shop, and the tattoo aficionado.

Ten Chapters   144 Pages $27.95   Over 400 photos, 100% color

Wolfgang Publication Titles

For a current list visit our website at www.wolfpub.com

ILLUSTRATED HISTORY

Ultimate Triumph Collection	$49.95

BIKER BASICS

Custom Bike Building Basics	$24.95
Custom Bike Building ADVANCED	$24.95
Sportster/Buell Engine Hop-Up Guide	$24.95
Sheet Metal Fabrication Basics	$24.95

COMPOSITE GARAGE

Composite Materials Handbook #1	$27.95
Composite Materials Handbook #2	$27.95
Composite Materials Handbook #3	$27.95

HOT ROD BASICS

Hot Rod Wiring	$27.95
How to Chop Tops	$24.95
How to Air Condition Your Hot Rod	$24.95

MOTORCYCLE RESTORATION SERIES

Triumph Restoration - Unit 650cc	$29.95
Triumph MC Restoration Pre-Unit	$29.95

CUSTOM BUILDER SERIES

How to Build A Café Racer	$27.95
Advanced Custom Motorcycle Wiring - Revised	$27.95
How to Build an Old Skool Bobber Sec Ed	$27.95
How To Build The Ultimate V-Twin Motorcycle	$24.95
Advanced Custom Motorcycle Assembly & Fabrication	$27.95
Advanced Custom Motorcycle Chassis	$27.95
How to Build a Cheap Chopper	$27.95
How to Build a Chopper	$27.95

SHEET METAL

Advanced Sheet Metal Fabrication	$27.95
Ultimate Sheet Metal Fabrication	$24.95
Sheet Metal Bible	$29.95

AIR SKOOL SKILLS

Airbrush Bible	$29.95
How Airbrushes Work	$24.95

PAINT EXPERT

How To Airbrush, Pinstripe & Goldleaf	$27.95
Kosmoski's New Kustom Painting Secrets	$27.95
Advanced Custom Motorcycle Painting	$27.95
Pro Pinstripe Techniques	$27.95
Advanced Pinstripe Art	$27.95

TATTOO U Series

Into The Skin The Ultimate Tattoo Sourcebook	$34.95
Tattoo Sketch Book	$32.95
American Tattoos	$27.95
Tattoo - From Idea to Ink	$27.95
Advanced Tattoo Art	$27.95
Tattoo Bible Book One	$27.95
Tattoo Bible Book Two	$27.95
Tattoo Bible Book Three	$27.95

NOTEWORTHY

American Police Motorcycles - Revised	$24.95
Guitar Building Basics Acoustic Assembly at Home	$27.95

LIFESTYLE

Bean're — Motorcycle Nomad	$18.95
The Colorful World of Tattoo Models	$34.95

References

Although much of this book was based upon the experiential knowledge of the author, the following sources helped provide supplemental information to the topics covered in this handbook:

www.fibreglast.com (Has videos on composite tooling fabrication, and information on pre-preg, cores, resin infusion, and a variety of composites materials.)

Gougeon, Meade. *Gougeon Brothers on Boat Construction: Wood and West System Materials, Fifth Edition.* Bay City, MI: West Publishing, 2005. (Although this book is geared mainly towards boat building, it is a great resource for a variety of molding questions. This same publisher also has good reference materials on vacuum bagging techniques, composite repair, and other helpful topics.)

Strong, A. Brent. *Fundamentals of Composites Manufacturing: Materials, Methods, and Applications, Second Edition.* 2007 (A good technical read about composites technologies and how to apply them through a variety of manufacturing methods.)

www.aircraftspruce.com (Supplier of composite materials, including a wide variety of resins, reinforcements, cores, etc.)

www.piercanusa.com (Supplier of latex bladders with good reference information about bladder and mold design.)

www.uscomposites.com (Supplier of several composite materials with good information about how to use these various materials.)

www.precisionboard.com (The website for Coastal Enterprises containing good information about a variety of foam products for moldmaking and cores.)

www.freemansupply.com (A great resource for resin and casting products, and general moldmaking and composites information.)

www.hexcel.com (Large supplier of reinforcements and cores, with excellent product and composite design information.)

www.plascore.com (Supplier of cores that furnishes good reference information about designing for cores and using infusion processes with them.)

SAMPE Journal and www.sampe.org (Materials sources from the Society for the Advancement of Material and Process Engineering. Contains excellent information on the use of composites through numerous journal articles.)

www.compositesworld.com (Publisher of *Composites Technology and High Performance Composites* magazines - a great resource for composites info.)

www.ingramcontent.com/pod-product-compliance
Lightning Source LLC
Chambersburg PA
CBHW061405160426

42812CB00088B/2574